Mammals: A Very Short Introduction

Very Short Introductions available now:

WATER John Finney
WEATHER Storm Dunlop
THE WELFARE STATE David Garland
WILLIAM SHAKESPEARE Stanley Wells
WITCHCRAFT Malcolm Gaskill
WITTGENSTEIN A. C. Grayling
WORK Stephen Fineman

WORLD MUSIC Philip Bohlman
THE WORLD TRADE
    ORGANIZATION Amrita Narlikar
WORLD WAR II Gerhard L. Weinberg
WRITING AND SCRIPT
    Andrew Robinson
ZIONISM Michael Stanislawski

Available soon:

ENVIRONMENTAL LAW
    Elizabeth Fisher
PERCEPTION Brian J. Rogers

PROJECTS Andrew Davies
SOUTHEAST ASIA James R. Rush
BIG DATA Dawn E. Holmes

For more information visit our website

www.oup.com/vsi/

T. S. Kemp

MAMMALS

A Very Short Introduction

OXFORD
UNIVERSITY PRESS

OXFORD

UNIVERSITY PRESS

Great Clarendon Street, Oxford, OX2 6DP,
United Kingdom

Oxford University Press is a department of the University of Oxford.
It furthers the University's objective of excellence in research, scholarship,
and education by publishing worldwide. Oxford is a registered trade mark of
Oxford University Press in the UK and in certain other countries

Published in the United States of America by Oxford University Press
198 Madison Avenue, New York, NY 10016, United States of America

British Library Cataloguing in Publication Data

Data available

Library of Congress Control Number: 2017938609

ISBN 978-0-19-876694-0

Printed in Great Britain by
Ashford Colour Press Ltd, Gosport, Hampshire

Contents

List of illustrations

Mammals

Chapter 1
What is a mammal?

We are mammals. So are horses, lions, moles, anteaters, a bumblebee bat weighing 2 grams, a blue whale weighing 100,000,000 grams, and many more. They are all mammals because they all share certain features that distinguish them from all other animals such as fish, reptiles, and birds. The females have mammary glands for feeding their young with milk and this gives them their name. Mammals also have a single bone in the lower jaw, three little sound-conducting ear bones, a very large forebrain, a permanently warm body, and a high level of energy expenditure. These and many other characteristics that they share tell us that the mammals all descended from a single common ancestor, and this is why they are put together into a formal Class Mammalia as one of the groups of the vertebrate animals.

That said, most but not all mammals share numerous other features. They give birth to live young rather than laying eggs (but not the duck-billed platypus or the echidnas); they are covered in a hairy pelt to reduce the loss of heat from the body (but not whales, pangolins, or humans); they have large molar teeth bearing several cusps for efficiently chewing their food (but not anteaters or whalebone whales); and they walk holding their four feet below the body (but not bats, moles, or gibbons). What the exceptions point to is the amazing variety of different kinds of mammals that have evolved. There are walkers, runners, hoppers, diggers, and burrowers; many are swimmers, tree dwellers, gliders, or fliers. There are those that feed on mammals larger than themselves, and those that live solely on ants and termites, or on fish. Every imaginable kind of vegetation is utilized by one mammal or another, from nutritious harvests of fruits and nuts, through foliage and tubers, to dry leaves, bark, and twigs. Furthermore, mammals live in virtually every part of the world. Arctic foxes survive in temperatures down to −70°C; camels, oryxes, and many rodent species exist in the hottest of deserts; while tropical forests and grasslands teem with species. Throughout the freshwater and marine realms, and in high snowy mountains, there are mammals making a successful living. What links them all is their fundamentally similar biology that we might name 'mammalness', which allowed the Class to diversify into such a great variety of forms, lifestyles, and habitats: we shall explore the nature of this biology shortly. First, however, we must quickly run through all the different groups and kinds of mammals that populate the Earth today.

The kinds of living mammals

The 5,500 or so species of mammals alive today fall into three very unequal groups (Figure 1). The monotremes (Monotremata) occur in Australasia, where they consist only of the duck-billed platypus (*Ornithorhynchus*; see Figure 26(b) later in the volume) and the spiny anteaters (*Tachyglossus* and *Zaglossus*; see Figure 25(b) later in the volume). These separated from the other mammals long ago and are the only mammals that still lay eggs rather than producing live young, although the embryo only stays in the egg for a very short time. After that, it emerges at an extremely immature, helpless stage to feed on milk from the mother's mammary glands just like other mammals. Apart from this, and one or two technical features of the skeleton, monotremes are no different from other mammals in their general biology. Both are specialized for their respective way of life: they both have short limbs extended sideways rather than downwards—which the platypus uses for swimming down to the bottom of lakes and rivers to collect invertebrates, and the echidnas to dig out ants and termites, and to burrow.

The second group are the marsupials found in Australasia, South America, plus a few in North America. They are so-named because the female has a pouch, or *marsupium*, on her abdomen, where she houses and protects her newborn offspring. Each growing infant stays attached to a teat, feeding on milk until it is well enough developed to emerge. There are only about 500 species of marsupials, constituting just 10 per cent of all living mammals. Nevertheless there are many different kinds. The kangaroos, koala, and bear-like wombats are medium- to large-sized herbivores with specialized grinding dentitions. The dasyuromorphs are carnivores and include the famous wolf-like thylacine, or Tasmanian tiger, which was hunted to extinction by the early 20th century. The last certain record of a living specimen was in Hobart Zoo in 1936—from time to time since then there have been reports of possible sightings, and even alleged photographs have been produced, but none have been very convincing. Among dasyuromorphs still alive today, there are the Tasmanian devil, a small hyaena-like animal noted for its ferocity, and quolls, which are rather cat-like. Most marsupial species are smaller bodied insectivores and omnivores, such as the rabbit-like bandicoots, and the opossums (called possums in Australia). Some of the Australian opossums live in the high forest canopy and are specialized gliders. They have a membrane between the four outstretched legs, which they can use to glide for up to 100 metres from one tree to another. The marsupial mole is another superbly adapted member of the group. It resembles a common mole in body shape, lack of ears, and short, stout limbs, and it survives in the Australian deserts by burrowing, and eating a diet of underground beetle larvae, other insects, and even small lizards when it comes across them.

The third and by far the largest mammalian group are the placentals. They differ from marsupials in the way their embryos

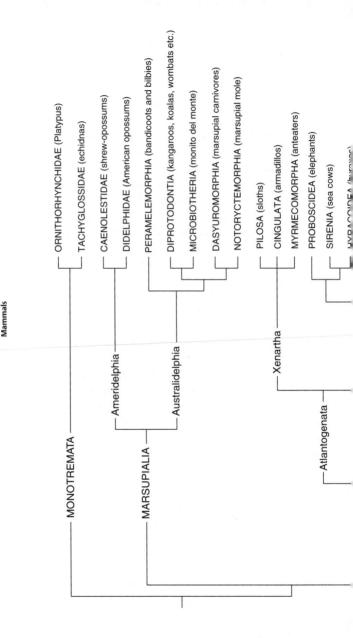

Mammals

- MONOTREMATA
  - ORNITHORHYNCHIDAE (Platypus)
  - TACHYGLOSSIDAE (echidnas)
- MARSUPIALIA
  - Ameridelphia
    - CAENOLESTIDAE (shrew-opossums)
    - DIDELPHIDAE (American opossums)
  - Australidelphia
    - PERAMELEMORPHIA (bandicoots and bilbies)
    - DIPROTODONTIA (kangaroos, koalas, wombats etc.)
    - MICROBIOTHERIA (monito del monte)
    - DASYUROMORPHIA (marsupial carnivores)
    - NOTORYCTEMORPHIA (marsupial mole)
- Atlantogenata
  - Xenartha
    - PILOSA (sloths)
    - CINGULATA (armadillos)
  - MYRMECOMORPHA (anteaters)
  - PROBOSCIDEA (elephants)
  - SIRENIA (sea cows)
  - HYRACOIDEA (hyraxes)

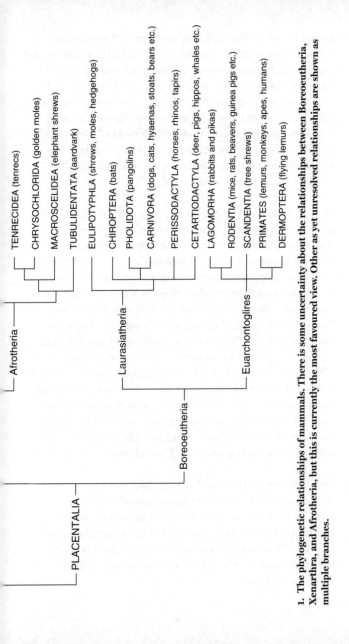

1. The phylogenetic relationships of mammals. There is some uncertainty about the relationships between Boreoeutheria, Xenarthra, and Afrotheria, but this is currently the most favoured view. Other as yet unresolved relationships are shown as multiple branches.

develop, the latter remaining for a much longer time within the mother's womb or uterus, where they are attached by a multi-layered placenta which is longer lasting than the much simpler, short-lived marsupial placenta. The young are therefore born at a much more developed stage, and they depend on the mother's milk for a shorter time before they are independent. The approximately 5,000 placental species are classified into twenty Orders of which almost 3,000 species are either rodents or bats—so if there is such a thing as a 'typical' mammal, it is a small, nocturnal, insectivorous, or omnivorous animal.

Until early this century, the evolutionary relationships of the Orders of placental mammals to one another were not at all clear. Since then, however, the technique of using gene sequences to discover relationships has led to a complete and widely accepted, though very surprising, *phylogenetic tree*. It turns out that there are three Superorders, associated respectively with the continents in which they first evolved. Each one is the result of a geographically separate evolutionary radiation long ago. The Afrotheria, as its name implies, is made up of several Orders whose earliest fossil record centres on Africa. They are the elephants (Proboscidea); the hyraxes, such as the rock dassies (Hyracoidea); the entirely marine dugongs and manatees (Sirenia); the shrew-like tenrecs of Madagascar (Tenrecidea); the long-snouted elephant shrews (Macroscelidea); the golden moles (Chrysochlorida); and the single, curious, ant and termite eating aardvark (Tubulidentata).

The smaller second Superorder, Xenarthra, consists exclusively of Southern and Central American mammals. These are the armadillos, sloths, and anteaters (such as the tamandua). In addition to their molecular similarities, xenarthrans share one very distinctive anatomical feature, which is that they have additional joints connecting the vertebrae in the vertebral column.

The third placental Superorder is the largest one, Boreoeutheria. Its name refers to the northern continents where it was originally based. At one time these continents made up a single land mass called Laurasia, but this has since broken up into North America, Greenland, and Eurasia. Most of the boreoeutherian groups have long ago expanded into the southern continents of Africa and South America, and some rodents and bats even managed to reach Australia. There are actually two branches of boreoeutherians. One of these is made up of the primates, rodents (Rodentia), rabbits (Lagomorpha), tree shrews (Scandentia), and the flying lemurs (Dermoptera). It has been given the awkward but logical name Superorder Euarchontoglires (Archonta was an older taxon for primates, tree shrews, colugos, and bats which is discredited because we now know that bats do not belong with the others; Glires is the name of the taxon that includes rodents and rabbits). The other boreoeutherian branch consists of several very different kinds of mammals that make up the Superorder Laurasiatheria. The least specialized of these are the small, insectivorous shrews, moles, and hedgehogs (Eulipotyphla); the most specialized are the bats (Chiroptera). Then there are medium- to large-sized carnivores such as dogs, cats, hyaenas, mongooses, bears, and pandas. Most of the world's medium to large herbivorous mammals are also laurasiatherians. They are the odd-toed ungulates (Perissodactyla) which have five, three, or single hoofed toes, such as the horses, rhinoceroses, and tapirs, and the even-toed ungulates (Artiodactyla), in which the feet have either four or two hoofs. This second Order is much the larger of the two and dominates the great grassy plains and much of the forests of the world. There are the antelopes mostly in Africa, and the deer in the northern continents, along with such familiar animals as pigs, sheep and goats, cattle, giraffes, camels and llamas, and hippopotamuses. Despite the huge anatomical differences, similarities in molecular sequences show that the Cetacea (whales and dolphins) are also unquestionably evolved from artiodactyls. Once this was discovered, an Order Cetartiodactyla had to be created to include both. In fact the closest living relatives of the whales turn out to be the hippopotamuses, which is interesting when we think that, despite the differences, both are large water-dwelling mammals. The final Laurasiatherian Order is also the oddest—it is the pangolins (Pholidota), powerfully clawed, ant eating mammals, made even more bizarre by a covering of protective horny scales instead of hair, and a tail by which it hangs from branches for safety.

Chapter 2
The biology of mammals

If we want to understand mammals and their amazing variety, we must first learn about the underlying biology that they all share. All the different kinds of mammals can then be explained by the different ways in which this general nature, this 'mammalness', became modified by evolution for the many different lifestyles and habitats.

How and why mammals stay warm and active

The key to understanding how mammals have adapted to lead such a variety of lifestyles in such an incredibly wide range of habitats is their *metabolic rate*. This is the measure of how rapidly the chemical processes in their cells can produce the energy needed for the animals' activities. These include muscle contraction during locomotion, the many internal bodily functions, and the processing of large amounts of sensory information in the brain. Energy is also needed as heat to stop the body from cooling down when conditions are cold. All this energy comes from the breakdown of carbohydrates and fats in the food to water and carbon dioxide. The metabolic rate of any individual animal ranges from the lowest, or *basal level*, when it is at rest to the maximum level when, for example, it is running as fast as it can. Both these measures, basal and maximum, vary greatly amongst different kinds of organisms. In the case of mammals, the metabolic rate is extraordinarily high—no less than five to ten times those of a lizard or crocodile, for example. There are two important reasons for the high metabolic rate. One is that it allows a mammal to keep up a much higher level of continuous activity such as running or fighting without getting out of breath and having to stop and rest: think of a pack of wolves chasing a herd of deer. The second is that the extra heat keeps the body at a higher, constant temperature, letting mammals live an active life over a wider range of external temperatures, such as during the night as well as the day, and in cold seasons as well as warm ones: think of nocturnal bats and arctic polar bears.

This form of temperature physiology involving a high metabolic rate, a high and constant body temperature, and a high maximum level of activity is termed *endothermy*, because the main heat source is the internal chemical reactions of the body tissues, and not the external environment as in *ectotherms* like amphibians and reptiles. Virtually every aspect of the life of mammals is linked one way or another to endothermy, either as part of the processes bringing it about or as part of the consequences. The high metabolic rate is due to the activity of the large number and size of the *mitochondria* in the cells. These are the sites of cellular respiration, and the heat they generate is distributed around the body by the blood stream to maintain the body temperature. The liver, kidneys, intestine, and brain generate most of the heat in a resting mammal, while the muscles produce the largest amounts when it is active.

The way the mammal's body temperature is kept constant is by using the large amount of heat produced in the tissues to raise the body temperature above the external temperature. This creates a temperature gradient so that heat flows away from the body. The rate at which heat is lost is finely controlled by changing, minute by minute, how much of the heat is allowed to be conducted away through the skin. The volume of warm blood flowing to the skin can be altered by opening or closing the skin capillaries.

The insulation offered by the pelt can be changed by fluffing up or flattening the hair, so that a thicker or a thinner layer of air is trapped. If the body temperature begins to rise too high, such as during exercise, the heat loss is increased; if it falls too low such as at night, the loss is reduced. However, there are limits to the range of external temperatures within which the mammal can control its body temperature by this means alone, and this is called the *thermo-neutral zone* (TNZ; Figure 2(a)). At an external temperature above the TNZ, the body cannot lose heat fast enough through the skin and the rate of heat loss must be increased by evaporation: of sweat spread over the skin from special sweat glands, or from the moist membranes of the mouth by shallow panting. On the other hand, if the ambient temperature falls below the TNZ, heat is lost too fast for the skin to be able to prevent the body from cooling down, and so heat production must increase. This may be by shivering, when the muscles generate extra heat or, on a longer time scale such as during a cold season, by increasing the basal metabolic rate. This way, all mammals can live and be active over a greater range of external temperatures than can ectothermic animals such as reptiles and amphibians, which rely on the sun to maintain their body temperature at the right level. Furthermore, different species of mammals have evolved different TNZs, to suit their particular habitats. At one extreme, the Arctic fox has such thick fur that the lower limit of its TNZ is about −40°C, a temperature it can therefore withstand without having to increase its metabolic rate at all. Below that its metabolic rate does increase, sufficiently to allow it to survive and remain active in temperatures as low as −70°C. However, it does face the opposite danger of overheating if it is too active, even at temperatures not much above freezing level. In contrast, typical small tropical mammals have a lower limit still well over 20°C. At temperatures below this, many have to enter into a temporary state of *torpor*, or suspended animation, because they can no longer prevent their body temperature from falling. On the other hand, they can be active with no need to waste water by evaporation at temperatures of well over 30°C. Longer term

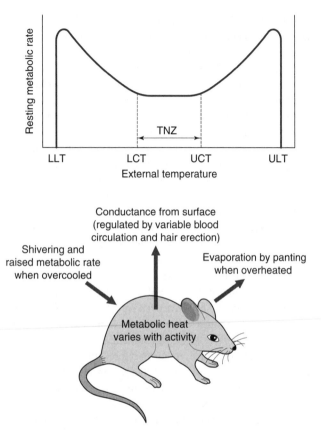

2. The relationship between the resting metabolic rate and the ambient temperature. Over the ambient temperature range of the thermoneutral zone (TNZ), the body temperature can be kept constant by varying the rate of heat loss through the surface alone. At an ambient temperature below the lower critical temperature (LCT), more metabolic heat must be produced. Above the upper critical temperature (UCT) more heat has to be lost by evaporation, in most mammals by panting moist air, and in some by sweating or covering the skin with urine. The body temperature can no longer be maintained below the lower lethal temperature (LLT) and either torpor or death by hypothermia ensues. Above the upper lethal temperature (ULT) death due to hyperthermia ensues.

torpor, or hibernation, is a solution to intolerably low seasonal temperatures in many mammals living in the temperate zones, such as European hedgehogs, bats, and grizzly bears.

Maintaining a high metabolic rate requires the intake of a lot of food, typically ten times more per day than that of a similar-sized ectotherm, which is why mammals spend most of their lives eating. Many different feeding strategies to achieve this have evolved in mammals. The uniquely mammalian kind of dentition, in which the individual teeth are specialized for preparing and masticating one or another type of food, makes digestion in the intestines easier and faster after it is swallowed. Herbivorous mammals also have an enlarged region of the intestine that acts as a *fermentation chamber*. This is full of microorganisms that can break down the cellulose and lignin of the plant cell walls, something mammals have never evolved their own enzymes for. Other remarkable specializations include the ability to eat fish with the help of a row of simple-pointed teeth and to consume ants and termites using a long, sticky tongue to collect and swallow them in sufficiently large numbers, with no need for teeth at all.

Another essential requirement for maintaining a high metabolic rate is the intake of large amounts of oxygen. When a mammal is not especially active, breathing is by expansion and contraction of the thorax using moveable ribs and their muscles, which draws air in and out of the lungs. However, during a bout of more intense activity, the maximum capacity of the lung is increased by means of the *diaphragm*, a curved sheet of muscular tissue bounding the hind wall of the thoracic cavity. When the diaphragm contracts, the volume of the thoracic cavity increases even more, increasing the volume of the gas flowing in and out of the lungs. Mammals also have a *secondary palate*, which is another adaptation to increase the overall rate of breathing. It is a sheet of bone lying in the roof of the mouth that separates the airway between the nostrils and lungs above it from the mouth cavity below it, so that food can carry on being chewed without having to interrupt breathing.

Blood circulation is modified by an ingenious device to increase the exchange of gases within the lungs and at the body tissues. The heart is divided into two sides, separated by a septum. Blood from the left hand side is sent exclusively to the lungs by a *pulmonary arch*, where it takes up oxygen and gives up carbon dioxide. This freshly oxygenated blood then returns to the right hand side of the heart, from where it is pumped again, this time to the body tissues, where it delivers oxygen and receives carbon dioxide. The effect of this *double-circulation* is twofold. The blood that is poorest in oxygen is sent to the lungs, while blood that is richest in oxygen is sent to the tissues. In this way the concentration gradients of the gas between blood capillaries and body cells is kept as high as possible at both sites. More oxygen is taken in at the lungs, and more is given up to the tissues than if the blood was all mixed up. Also, by pumping the blood twice instead of only once for each full circuit of the body, the average blood pressure is increased which creates a higher rate of blood flow.

The idea of endothermy is quite straightforward: simply increase or decrease heat loss as necessary to keep the body temperature constant. But to be effective there has to be a whole range of mechanisms to detect any change in body temperature and initiate suitable responses to correct it. To complicate matters even more, the system has to operate over different time scales. Sometimes it must react almost instantaneously, such as when a mouse spots a predator and rushes to its burrow, increasing its heat production. Temperature sensing cells in the *hypothalamus* at the base of the brain monitor the blood temperature. If this is too high, then a signal is sent via the nervous system to the skin to trigger an increase in heat loss by increasing blood flow to the skin and flattening the hairs on the skin. Conversely, if the blood temperature is too low, a signal is sent to the skin to conduct less heat by reducing the rate of blood flow and fluffing up the hairs.

The system also has to operate on the longer time scale of the regular diurnal changes in the external temperature, by initiating behavioural responses in the individual, such as to seek shelter, water, or shade from the sun, or to huddle together with other individuals to keep warmer. In smaller mammals, such as bats and many desert mice, where a temporary state of torpor can be induced when the body temperature is too low, the metabolic rate is allowed to fall to a fraction of its normal rate, with their breathing and heart rates slowing to about one breath or beat per minute, respectively, and their body temperature being kept only just above that of the outside temperature. Coming out of torpor is achieved by about 20 minutes of very energetic shivering of the muscles to raise the body temperature to normal. On a seasonal scale, the onset of winter conditions triggers many mammals to increase their basal metabolic rate using a special tissue called *brown fat* that is laid down around the organs, and whose sole function is to generate heat. Hibernation, which is longer term torpor, is a more extreme strategy used especially by smaller bodied, temperate zone species, such as ground squirrels, which are more vulnerable to heat loss than large bodied forms.

Endothermy has a serious consequence for the development of offspring. The smaller a mammal is, the more rapidly it loses its heat because of the higher surface area compared to volume, a simple law of physics. An endotherm cannot succeed if it is too small, which creates a problem for the early stages of the developing young. The solution is for the mother to provide an environment in which she controls the temperature and provides all the necessary nutrients and gases, within which the embryo can safely live. *Viviparity* is retaining the embryo in the uterus so that it can thrive and develop without requiring an ability to regulate its own temperature. Even so, the newborn young of many mammals such as mice and carnivores are still very small and immature, and they continue to need a controlled environment. This is now provided in the form of a nest or burrow created by the parent. In other mammals, notably large herbivores and whales, the embryos continue their development within the uterus to a far more mature stage, by which time they have a large enough body size, and their own regulatory system working to maintain their body temperature independently.

Like heating a house, endothermy is very costly in fuel, in the form of a high daily food requirement, and it is also very costly to evolve, because of all the integrated components that are necessary. Therefore we can be sure that there must be an adaptive benefit for mammals at least equal to these costs—otherwise endothermy would not have evolved at all. And there are indeed several benefits.

The first benefit is that the constant body temperature allows mammals to remain fully active, such as when feeding, hunting, and seeking mates, over a longer period of time on both a daily and a seasonal basis. Unlike ectotherms, mammals can be just as active during the night as during the day, and the majority of species, especially the small ones, are mainly nocturnal foragers. Furthermore, most mammals can continue to lead their life normally throughout the seasons, cool and warm. Only in higher latitudes do many of the smaller and a few of the larger species resort to hibernation in the coldest season. This is why mammals can occupy such a wide range of habitats globally, on land, in freshwater, and in the sea.

A second benefit of the constant body temperature relates to the internal workings of the body. Most of the chemical and physical processes in the body are sensitive to temperature, including the rates of the thousands of enzyme controlled chemical reactions; the rate of diffusion of the molecules that transmit messages from nerve cell to nerve cell; the rate of contraction of muscles; and the viscosity of blood and therefore its rate of flow through the blood vessels. All organisms are highly *integrated systems*, consisting of a great many such processes linked to one another. Outside quite narrow limits, changes in temperature affect the different processes enough for this integrated activity to break down, at which point the well-organized life of the organism starts to deteriorate. Generally, the greater the complexity of the organism, the narrower the level of change in body temperature that can be tolerated. Thanks to the constancy of their body temperature, mammals are more complex than ectotherms, in other words they have a greater number of interacting integrated processes. The brain is the most important part of a mammal to which this applies, because it consists of huge numbers of interacting nerve cells. The endothermic brain is up to ten times the volume of that of a similar-sized reptile. Its 10^9–10^{10} neurons can receive vastly larger amounts of different sensory information, integrate this with greater levels of learning and cognition, and generate a wider range of accurately controlled behavioural outputs. It is also no coincidence that when body temperature regulation fails and a mammal fatally overheats or cools down, the actual cause of death is invariably the failure of some part of its brain function, such as loss of control of respiration, heart failure, or coma.

In addition to the benefits of a constant body temperature, endothermy endows mammals with a higher level of sustainable activity, known as the *maximum aerobic metabolic rate* (MAMR). Whilst it is the mitochondria of the cells of the internal organs that are mainly responsible for the high basal metabolic rate and hence the elevated constant body temperature, the muscles also have large numbers of mitochondria. Their primary role is to break down glucose by oxidation to provide the mechanical energy for muscle contraction, and having a large number of mitochondria means that more glucose can be metabolized and energy supplied at a higher rate. The outcome is that the MAMR of a typical mammal is around ten times that of a reptile of the same body weight. The two can actually run at broadly the same maximum speed. The difference is that after a mere two or three minutes the reptile accumulates an oxygen debt ('out of breath') so large that it has to stop and recover, while the mammal can continue indefinitely at that speed, until its food reserves run out. There are numerous behavioural and ecological advantages to this endurance when hunting, foraging, evading predators, migrating to new seasonal food sources, and finding a mate, even though it is so costly in terms of nutritional requirements.

Taking together the extraordinarily wide range of environmental conditions and ways of life that it opens up, endothermy can indeed be fairly said to capture the essence of what it means to be a mammal—of 'mammalness'.

How mammals save water

Osmoregulation is maintaining the correct concentration of water, or osmotic pressure, of the body fluids, and is just as important for the proper functioning of the body as regulating the temperature. As humans sometimes find out to their cost, often the most severe problem facing an animal living on dry land is loss of water due to a combination of the moisture in the breath, expulsion of waste products as urine and faeces, and cooling by evaporation. It causes the osmotic pressure to increase, which, if not corrected, has potentially disastrous consequences for the animal's metabolic processes.

On its own, the simple solution of drinking water fast enough to replace large losses would restrict life too much for mammals because they would have to spend their whole time within easy reach of a freshwater source, as frogs and other amphibians do. They have therefore evolved several ways of reducing water loss in the first place. First, the skin is waterproofed by having a layer of the dry protein keratin, which is also important because it is very tough and protective. Second, the faeces are dried as much as possible by reabsorption of water in the rectum. Third, mammalian kidneys, unlike those of any other animal, can produce urine that is very highly concentrated. Each of the many fine tubules in the kidney has a long, U-shaped addition called the *loop of Henle* which is surrounded by capillaries (Figure 3). Sensory cells in the

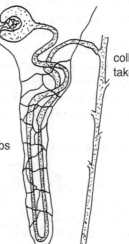

glomerulus - ultrafiltrates
the blood plasma

collecting duct
takes urine away

loop of Henle reabsorbs
water molecules to
concentrate the urine

3. A single mammalian kidney tubule and its capillaries.

brain measure the osmotic pressure of the blood, and a message
via a hormone is sent from the pituitary gland at the base of the
brain to the loop of Henle. Here it stimulates the loop to reabsorb
the right amount of water back into the blood to keep the
concentration of the blood at the correct level. These ways of
saving water are so effective that many mammal species living in
arid conditions where water is scarce never need to drink at all,
but get enough water from their food alone. For example, the
kidney of the kangaroo rat of North American deserts can produce
urine that is sixteen times as concentrated as its blood. Compare
this with our human kidney, which can only produce urine four
times as concentrated as our blood.

Teeth and guts: how mammals get their food

Mammals have specialized to feed on a huge variety of different
foods, from ants to antelopes, grass seeds to tree bark. This

variation is due to modifications via evolution of an ancestral style of feeding that is still found today in many small mammals, such as that of opossums amongst the marsupials, and that of shrews, tree shrews, and tenrecs amongst the placentals. The dentition of these generalized mammals consists of four different kinds of teeth (Figure 4), each performing a different function. At the front of the jaws the simple, pointed *incisor* teeth are for picking up small food items such as insects or seeds, and they are often useful for grooming the fur as well. Behind the incisors there is a single

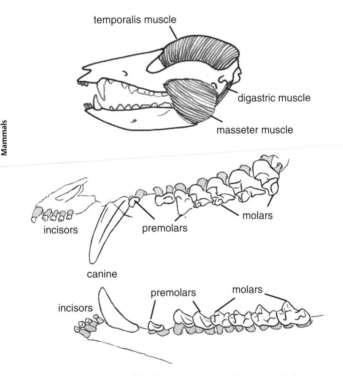

4. Feeding mechanism of the Virginia opossum: (top) the skull showing the three main muscles, temporalis and masseter for jaw closing and digastric for jaw opening; (below) the upper and the lower right tooth rows as if seen from the inside of the mouth.

canine tooth in the upper and the lower jaw which is larger than the incisors, and whose function is to disable small items of living prey, and in many mammals it is also used for other functions, such as scratching around the ground, fighting, and defence. The remaining kinds of teeth are unique to mammals, being enlarged and bearing more than a single cusp on the crown. The three or four *premolars* only have one or two extra cusps and are used for simple crushing and breaking up of food in the mouth. The three or four *molars* at the back are the largest teeth, and each one possesses six or more cusps connected by sharp ridges. During chewing, the lower molars meet the upper molars in a precise fashion called *occlusion*. The lower ridges slide against the upper ridges like the blades of tiny pairs of scissors, performing a cutting action. The food in between is reduced to a fine pulp ready for swallowing, now in a state that can be digested and absorbed in the intestine much more quickly. Teeth occluding in this way need a large bite force to produce a high enough pressure between them to cut up the food. At the same time the movement of the lower jaw must be very precise to make sure that the lower teeth accurately meet the upper teeth. The biting muscles between the cranium and the lower jaw (Figure 4) achieve both these requirements. They are very large, which provides the required pressure. They also consist of an inner (*temporalis*) and an outer (*masseter*) part, so that the jaw is held in a muscular sling allowing finely controlled movements of the jaws in the fore-and-aft and side-to-side directions.

This differentiation of the teeth, along with the forcefulness and accuracy of jaw closing, is unique to mammals, and in its basic form it is perfect for the intake of a diet of small items of nutritious food. It has also proved to be amazingly adaptable, evolving to suit the wide range of specialized diets consumed by different mammalian groups, as will be seen in later chapters. Generally speaking, the evolution of the teeth and jaws of carnivorous mammals consisted of enlargement of the canines, for killing their prey, and of the cutting rather than the crushing parts of the molar teeth, to slice the flesh into easily swallowed pieces. Herbivores, on the other hand, evolved an exaggerated crushing action of the premolars and molars, and more extensive horizontal jaw movements for finely grinding their food. Fish eaters evolved simple, sharp, pointed teeth suitable for spearing fish, with weak jaw muscles because their food could be easily swallowed whole. Some mammals have evolved very reduced or even a complete loss of teeth, notably the various kinds of ant and termite eaters, which use a long, sticky tongue to collect the food, and the baleen whales, which filter plankton directly from huge mouthfuls of sea water.

Locomotion: how mammals get from place to place

Just as adaptation for different diets reflects the great variety of ways of life of mammals, so too do their means of locomotion. Fast running, jumping, burrowing, tree climbing, flying, and swimming all evolved from an ancestral mode of locomotion that is still seen today in small, non-specialized mammals such as tree shrews, hedgehogs, and opossums.

Compared to typical reptiles, such as lizards, whose limbs sprawl out sideways, the mammalian elbow is turned backwards and the knee turned forwards, towards the body, so that the feet lie more or less underneath (Figure 5). The left and right feet are therefore quite close together and this makes a mammal less stable but correspondingly more agile, rather like a motor cycle compared to a car, making the mammal much more effective at activities like accelerating, turning sharply, and scrambling over rough terrain. A second effect of bringing the legs under the body is that it raises the rib cage above the ground and makes it easier for the animal to carry on breathing, whether stationary or moving.

The spinal column of a mammal consists of several regions, each having a different function. The first region is the neck (*cervical*),

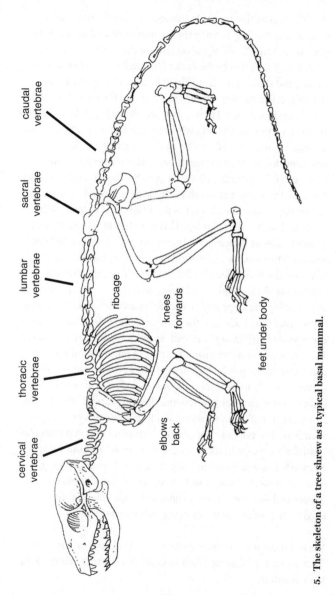

caudal vertebrae

sacral vertebrae

lumbar vertebrae

thoracic vertebrae

cervical vertebrae

ribcage

knees forwards

feet under body

elbows back

5. The skeleton of a tree shrew as a typical basal mammal.

which is responsible for the extreme mobility of a mammal's head, aiding feeding, sensing the environment, and much of its other behaviour. Nodding movements of the head occur at the specialized joint between the back of the skull and the first vertebra (*atlas*), and shaking the head about its axis takes place at the equally specialized joint between the atlas and the second vertebra (*axis*). Bending the head from side to side is the job of the remaining five neck vertebrae. The following thirteen vertebrae make up the *thoracic* region. They all carry a pair of long ribs attached by a moveable double head in such a way that the ribs can rotate forwards and outwards to increase, and backwards and inwards to decrease, the volume of the thoracic cavity during breathing. The second function of the rib cage is for attachment of the shoulder girdle and with it the front leg. This is quite a loose connection, using muscles and connective tissue rather than a bone to bone contact. The mobility of the shoulder girdle on the rib cage increases the overall length of the fore limbs' stride. Five stout *lumbar* vertebrae, lacking ribs, come next. They are responsible for transmitting the large walking forces generated by the back legs to the rest of the body, and they also act as a girder to carry the considerable weight of the intestines, liver, bladder, and kidneys. The five *sacral* vertebrae form the part of the vertebral column connected to the pelvis. This needs to be a strong connection of bone directly to bone, because it is the place at which the walking force generated by the hind limbs is applied. Finally, the mammalian tail (*caudal*) region is a flexible string of small, weak vertebrae. The tail does not have an important role in mammalian locomotion except in some specialized forms, for example as a counterbalance in jumping kangaroos, a prehensile grasping organ in spider monkeys, and a support for the tail flukes of whales and dugongs. In the majority of mammals it is only used for activities like fly-swatting, and social signalling.

We will meet the various ways in which this ancestral mode of locomotion has been modified by evolution in other mammals in later chapters.

Sense organs: a mammal's picture of its environment

As to be expected of animals that are as responsive to their immediate environment as mammals, their sense organs give them a great deal of useful information. In most mammals *olfaction*, the sense of smell, is important to an extent and degree of sensitivity impossible for humans to appreciate because our own sense of smell is so poorly developed. It is fair to say that in the absence of such a rich olfactory experience, the nocturnal life led by so many, especially small, mammals would be impossible. For example, the receptors of a rat's olfactory organ in its nose can detect over 1,000 different air borne molecules, called *odorants*, at such incredibly low concentrations that even a single molecule may stimulate a receptor cell. Equally impressive is the extent of the integration within the brain of the olfactory information received by the olfactory organ. The nerve impulses from the receptor cells for each odorant travel to small neural units called *glomeruli* in the olfactory bulb at the front of the brain, which are spatially arranged to create a sort of 'olfactory map'. From this pattern, higher order nerve cells detect combinations and intensities of individual odorants that correspond to particular smells in the habitat. The result is comparable to the richness of the visual map with which humans are familiar. Projection of this information to the olfactory region of the cerebral cortex of the brain leads to further processing, and integration with other kinds of sensory information. There is a second organ sensitive to smell inside the nose, called Jacobson's organ, which is for detecting *pheromones*. These are molecules exuded by individuals and play a major role in communication and social behaviour between members of a group. An individual's social position in the community, sexual condition, aggressiveness, and maternal behaviour may all be communicated in this way.

The sense of hearing is also remarkably well-developed in mammals, and extremely so in certain groups that emit ultra-high frequency sounds for *echolocation*. A few species of shrews and other small insectivorous mammals use the echoes of sound frequencies not much above the human hearing limit of 20 kHz to investigate the coarser features of their immediate surroundings. Bats and whales use very much higher frequencies, up to160 kHz in dolphins and as much as 200 kHz in some bats, to provide a very detailed picture of their nocturnal or marine habitat, where vision is of little use. There is a chain of three tiny *ear ossicles* in the mammalian middle ear called the hammer (*malleus*), the anvil (*incus*), and the *stapes*. They connect the ear drum, which vibrates when sound waves hit it, to a hole in the braincase leading to the *cochlea* that houses the actual sense organ. The role of the ossicles is to provide a lever effect that improves the impedance matching between the air that carries the sound waves falling onto the ear drum and the fluid that fills the cochlea. The cochlea itself is a narrow, coiled tube within the bony wall of the braincase and the sound sensitive *organ of Corti* extends along its length. Different parts of the organ of Corti are sensitive to different frequencies of sound, and because it is so elongated, mammals possess a very fine ability to discriminate between frequencies over a wide range: just think of even our ability to hear from the lowest bassoon to the highest piccolo notes. The external ear (*pinna*) is another uniquely mammalian device for improving the sense of hearing, acting like an ear trumpet to concentrate the sound collected. Furthermore, most mammals have mobile pinnae which help to detect the direction the sound is coming from. The ability to detect high frequency sound has also led most mammals to evolve high frequency vocalization, and verbal communication has therefore become another important part of their social behaviour.

Vision is a much less important sense than either smell or hearing in nocturnal mammals, which includes the majority of species. Their visual acuity is only 1–10 cpd (cycles per degree of visual arc; a measure of how well the eye can discriminate between closely spaced lines), with microbats representing those with the lowest, and larger predators and herbivores tending to have the highest acuity. Colour vision is also poor to non-existent in most mammals and there is a bichromatic system that at best can only discriminate blue, yellow, and green, but not orange and red. Contrast this with those species that are largely or exclusively diurnal, namely the higher primates, including ourselves, in which vision is by far the most important means of sensing the environment. Visual acuity is about 50 cpd, thanks to a special structure in the retina called the *fovea*, the point where incoming light rays are focused, and where the density of receptor cells is particularly high. Primates have also evolved trichromatic vision, with three sets of visual cones in the retina, making us sensitive to the full visual spectrum from red to violet.

Reproduction: few but well-cared-for offspring

Apart from the peculiar monotremes which never lost the ancient egg-laying habit of ancestral mammals, all modern mammals are *viviparous*, bearing their young live. The embryo attaches to the wall of the uterus by a placenta, across which nutritional molecules, respiratory gases, and waste products diffuse between the mother's tissues and the embryo's developing blood system. Here the developing foetus is physically protected, and by staying warm and well-nourished it can grow rapidly. The marsupial placenta is a simple structure consisting only of contact between the *yolk sac* of the foetus and the wall of the uterus. Diffusion takes place between the foetal and the maternal blood capillaries, and the time spent in the uterus, the *gestation* period, is relatively brief, typically no more than about two weeks. The young is born at an extremely immature stage, and scarcely any more than the olfactory organ, the mouth, and the clawed forelimbs are distinct. These, however, are its three most important parts. Immediately after it has emerged from the birth canal, its fore limbs allow it to cling to the mother's fur and using its olfactory organ it can follow a scent trail leading into the pouch. Once there, its mouth grasps a teat and permanently seals around it. The lactation period is much longer than was the gestation period, and the young continues to develop in the pouch until it reaches a stage where it is capable of independent life. The milk produced by the mother's mammary glands is a mixture of all the proteins, fats, carbohydrates, vitamins, and minerals necessary for growth, and it also contains antibodies produced by the mother that are important for resisting disease in the young. The composition of the milk varies over time, with more dilute milk early on, followed by a richer, higher protein-content fluid as the growth rate increases.

Placental mammals differ from marsupials in having a more complex, longer lasting placenta that provides the embryo with a more extended gestation period. The embryo quickly becomes deeply embedded in the wall of the uterus, and numerous fine, finger-like processes develop between the two, hugely increasing the surface area of contact for diffusion. The provision of nutrients and oxygen to the foetus is much greater than in marsupials, and this means it grows and develops faster. The outcome is that a placental infant is born at a far more advanced stage than a marsupial one and needs a relatively shorter lactation period. In large herbivores like cattle and horses, and also in whales, the newborn infant is already capable of a largely independent life, and can keep up with the group within an hour or so of birth. Another way that this placenta differs from the marsupial's is in secreting molecules that prevent the mother's antibodies from attacking the foetus. The foetus is not genetically identical to the mother and without this protection it would act like a foreign tissue graft and be rejected. This may be why placental embryos can spend so much more time in the uterus than marsupials. Placental milk is similar in composition to marsupial milk, although variable between species. For instance, the milk of marine mammals and of species living in cold areas such as reindeer has a much higher fat content than, say, that of humans and cows.

Brain and behaviour

If we really want to name the single thing that makes mammals so special and so successful and diverse in the modern world, we need look no further than the brain. An average mammalian brain is around ten times the size of that of a similar-sized reptile, and this is the external indication of a vastly greater number of neurons and interconnections between them. The large size of the brain is mainly due to an extensive, multi-layered structure called the *neocortex*, which dwarfs the remnant of the original forebrain of the ancestor. In mammals such as shrews and marsupials which have a less evolved brain, the neocortex is smooth, but in other taxa it has a pattern of grooves, called *sulcae*, which increase the surface area of the layers (Figure 6) and the number of neurons. All the information from the sense organs is relayed to the neocortex, which contains numerous interconnected *association centres*. Here sensory information about the environment is integrated with information received about the current state of the body, such as hunger, sexual condition, and body temperature, and also with associations accumulated during the animal's lifetime as a result of experience and learning. Using all this information, neural messages are generated, and pass via the brain stem and the nerve cord to elicit appropriate behavioural responses by the animal. The far greater amount of information that can be computed in this way in the mammalian brain, plus the huge variation in the possible responses it can initiate, explains why a mammal's reactions to its physical and social circumstances are so variable and nuanced.

Although the detailed patterns of behaviour vary widely amongst mammal species, there are several generalizations common to all. One is that they indulge in extensive exploratory behaviour to familiarize themselves with their particular locality, quickly learning the likely sources of food, and the safe spots for avoiding predators and rearing offspring. A second is the adaptable way

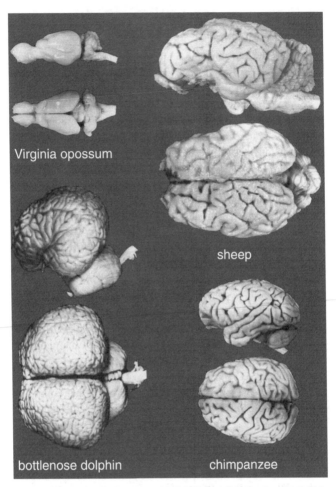

Virginia opossum

sheep

bottlenose dolphin

chimpanzee

6. Photographs of the lateral and dorsal views of the brains of Virginia opossum, sheep, bottlenose dolphin, and chimpanzee.

in which an individual can modify a general behavioural aim to suit the particular conditions it finds itself in. For example, most small mammals such as mice create some form of nest, burrow, or den in which to protect their young, but where to build it depends on finding a suitable place judged to provide a low risk of predation, and what to build it of depends on whatever material is available.

A third pattern of behaviour is the extensive range of signals that individuals within a species use to communicate with one another. In zebras, as many as twenty to forty different visual displays of facial expression and body posture have been noted, which are used to convey subtle shades of meaning by variations in the intensity, duration, or context of the signal. They act as close range signals to other members of the group, providing information about such things as mood, intentions, social status, aggression level, and reproductive condition. All mammals use secreted pheromones to affect the actions of others, and these can even alter the physiological state of the recipient. Some are applied as direct communication, individual to individual; some indirectly by scent marking the territory. Vocal signalling is important at night, over distances, and in dense foliage, and includes warning sounds to the group about the presence of a predator. A variety of high frequency noises, such as hissing, squeaking, and clicking of the teeth and lips, is often used to convey information about an individual's identity and social status, and to assess distances between group members. Pack hunting predators like wolves and the painted dogs of Africa use a range of sounds in order to coordinate the chase.

Many people have tried to correlate the patterns of mammalian social behaviour with particular habitats and modes of life, but there seem to be so many variables involved that we have as yet no clear picture. Social grouping varies from the solitary existence of the giant panda and aardvark, to the colossal herds of wildebeest and zebra migrating across the African savannah and Kalahari.

Within a group, there can be a mixture of males and females that are more or less equal, as with some lemurs, or at the other extreme a strict linear dominance from the alpha male downwards, as found with baboons. An elephant herd consists of a matriarch with her female descendants; the male offspring remain only until adolescence when they move away. In mice there is usually a single dominant individual but with the rest of the group being equal to one another. One unusual case is the naked mole rats, which have a complex social organization underground, with different roles for different individuals. A single reproductive queen is cared for, guarded, and fed by non-reproductive individuals.

Mating patterns vary too. The commonest system in mammals is *polygyny*, in which a male has several mates and plays no direct role in the rearing of the offspring. Other species, however, are *monogamous*, for example marmosets and many canids such as the wolf. Here a single male and female mate, often for life, and both parents provide care of the offspring. Rare but not unknown is *polyandry*, in which a single female has several mates and the males help care for the juveniles. The African painted dog often, though not always, behaves in this way. Finally, promiscuity occurs in some species, such as the prairie dog, in which both males and females have multiple partners within the colony.

The life history of mammals such as the size of the litter is another reproductive variable that is hard to relate to particular modes of life, mainly because it involves trade-offs between several considerations. Producing a larger litter has an obvious advantage to a female under favourable circumstances. However, the costs may include a reduced chance of survival of each individual offspring, especially if conditions such as the food supply deteriorate; and a possible reduction in the mother's future reproductive ability. Different mammals not only produce different litter sizes, but also vary in how mature the newborn offspring are. Some newborn mammals are extremely immature, such as marsupials, rats, and dogs, while others are developed enough to run with the herd shortly after birth, such as antelopes, cattle, and elephants and also some smaller species like guinea pigs. One factor to consider is that gestation only costs the mother about half as much in food compared to using lactation as a way to nourish the offspring, so a delay in birth represents a gain for the mother. But the counteracting costs are the greater vulnerability of the mother while pregnant, and the smaller number of offspring she could bear over her lifetime if birth is delayed.

Before reviewing the variety of mammals alive today, in Chapters 3 and 4 we shall take a look at their evolutionary history. We are fortunate enough to be helped in this by an exceptionally good fossil record, from the earliest appearance of animals with any mammalian characters at all, to the final establishment of today's mammalian fauna 320 million years later.

Chapter 3
The origin of mammals

We have a very good idea about what the ancestor of all the
mammals was like, and when it lived. Palaeontologists have
discovered the fossilized remains of small, rather shrew-like
animals in certain rocks that are about 200 million years old
(see Figure 10(d) later in the volume). These animals had all the
characteristics of the mammalian skeleton and teeth that were
described in the last chapter, although they do not belong to any
particular modern group. They are the earliest mammals known
and although not the actual ancestral mammal group, they were
certainly very similar to it. From what did they themselves evolve?
To answer this question we shall look at the fossil record of a group
of 'pre-mammals' that illustrate intermediate stages in the evolving
series through time, from the distant ancestor that mammals
shared with reptiles, to our first recorded mammal. They are
the *Synapsida*, sometimes called the 'mammal-like reptiles'.

Pelycosaurs

The story of the synapsids begins 320 million years ago in the
Upper Carboniferous Period with the *pelycosaurs* (Figure 7). These
heavily built animals were very unlike mammals, having simple
teeth, short legs sprawling outwards like a reptile, and a very
small brain. But they did have a few characteristics that link them

to mammals. One of these was the presence of a window in the skull bones behind the eye socket, called the *temporal fenestra*, for attaching stronger jaw muscles; and a second was a small canine tooth in each jaw. Pelycosaurs were restricted to the warm, permanently humid equatorial zone in the middle of what was at that time the single world supercontinent of Pangaea. It tells us that at this early stage synapsids were probably incapable of coping with the lower temperatures and periods of aridity they would have had to face in the seasonal climates of higher latitudes. Nevertheless, within this limitation pelycosaurs soon became the dominant land vertebrates of their day. They evolved into several different kinds of animal, and several grew to the considerable body size of 3–4 metres long. The ophiacodontids (Figure 7(c)) had long, narrow skulls and jaws bearing small, sharp teeth for catching fish, like a crocodile. Others, such as the edaphosaurs (Figure 7(b)), evolved shorter, stronger jaws with blunt teeth for a diet of plants. *Dimetrodon* (Figure 7(a)) was a large, ferocious predator with substantial canine teeth and a row of sharp-edged teeth behind them for ripping up its prey.

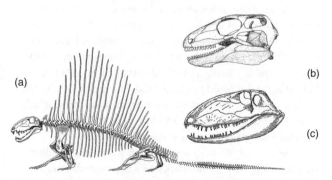

7. Pelycosaur-grade synapsids: (a) the sphenacodontid *Dimetrodon*; (b) the specialist herbivore *Edaphosaurus*; (c) the specialist fish eater *Ophiacodon*.

Therapsids

About 260 million years ago, by the Middle Permian, the world had grown increasingly arid, and the pelycosaurs had all but disappeared. This might have marked the end of the synapsid story, but fortunately a more advanced group called the *therapsids* had already evolved from pelycosaurs. The importance of this new stage is shown by the fact that therapsids are found in temperate latitudes, where the climate was distinctly seasonal, with both cooler and drier periods. They must have evolved to some extent the mammalian ability to be active over a wider range of temperatures and humidity levels.

The anatomy of the therapsids also supports the idea that they had evolved a higher metabolic rate and level of activity than pelycosaurs. *Biarmosuchus* (Figure 8(a)) illustrates the basic structure of the therapsids. It was about the size of a largish dog, and compared to its pelycosaur ancestors the skull was more strongly built and the temporal fenestra considerably larger. This shows that it had a stronger bite than its ancestors. The dentition was differentiated into robust conical incisors at the front, a very large upper and lower canine, and teeth behind the canines bearing fine serrations on the back edge. Clearly *Biarmosuchus* was a carnivore capable of feeding on relatively large prey. Its limbs were long and slender, and the shoulder and hip joints more flexible than in pelycosaurs, all indicating that it could run faster and with greater agility than the latter. Another important new mammalian characteristic is mammal-like bone tissue, which is an indication of rapid growth. Adding together the geographical evidence of distribution and the anatomical evidence for a high feeding rate, more effective locomotion, and a high growth rate, it is hard to avoid concluding that the therapsid metabolic rate was raised, though not to the extent that occurred in later stages of synapsid evolution.

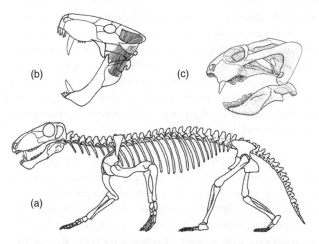

8. Therapsid-grade synapsids: (a) composite basal therapsid based on *Biarmosuchus* skull and gorgonopsian postcranial skeleton; (b) gorgopsian skull, a specialist hypercarnivore; (c) dicynodont skull, a specialist herbivore.

By Upper Permian times (265–250 million years ago), several new groups of therapsids had evolved and become by far the most abundant land vertebrates, just as the pelycosaurs had been before them. The gorgonopsians (Figure 8(b)) were medium to large bodied, agile carnivores armed with huge canine teeth, and with sharp, serrated incisors at the front of the jaws for ripping off pieces of meat for swallowing. Their main prey was a group of herbivores called the dicynodonts (Figure 8(c)). Many dicynodonts possessed a pair of large upper tusks (from which their name is derived), but the rest of their teeth were replaced by horny tooth pads on the jaws, much like turtles have today. These were used for slicing or crushing plant food using huge jaw-closing muscles. The fossils of well over a hundred species of dicynodonts have been found, from mouse- to rhinoceros-sized, and some of them were extremely abundant. We can compare this dominance of dicynodonts with the later reigns of the herbivorous dinosaurs in the Mesozoic, and the herbivorous mammals now. There were several other therapsid groups specialized in different ways. Dinocephalians were massively built, lumbering creatures as big as rhinoceroses, with the bones of their heads as much as 5 cm thick, perhaps for head-butting behaviour like that of bighorn sheep. The Therocephalia were slender-limbed, mostly cat-sized carnivores that had a long row of sharp, pointed teeth for living on a diet of invertebrates.

Fossil collecting is always partly a matter of luck, and while on an expedition in Zambia many years ago I was fortunate enough to spot a beautiful skeleton of an otter-sized animal called *Procynosuchus* (Figure 9(a)), with its little toothy jaws seeming to smile at me from the rock in which it was embedded. It is a primitive cynodont, the next important stage in our story of the origin of the mammals. The dentition of *Procynosuchus* was essentially mammalian. Like most other therapsids, there were sharp, pointed incisors at the front, followed by a larger canine tooth. But unlike them, the first five teeth immediately behind the canine were slightly expanded, whilst the last eight each had a ring of small extra cusps around the base. For the very first time, multi-cusped molar teeth had evolved, although they were very simple at this stage. The lower molars worked against uppers to chew the food in a crude way, so that when it was swallowed it could be digested and absorbed more quickly. Along with this new kind of dentition, the muscles that closed the jaws were enlarged and spread onto the outer as well as the inner side of the lower jaw. They could produce a bite force great enough for effective chewing. The pair of temporal fenestrae were greatly enlarged to accommodate these larger muscles, and now almost met at the midline of the skull behind the eyes. They were separated from one another by a narrow girder called the *sagittal crest*, which is still found in mammals. Another important new mammalian feature of *Procynosuchus* was a *secondary palate*, a sheet of bone in the roof of the mouth that separates food being chewed in the mouth from the air flowing from the nostrils to the lungs and back.

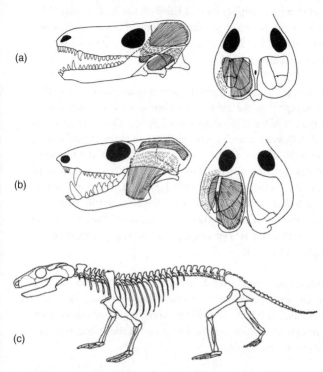

9. (a) *Procynosuchus* showing the start of the evolution of the
mammalian pattern of dentition and jaw musculature; (b) *Chiniquodon*,
showing the advanced cynodont arrangement; (c) skeleton of an
advanced cynodont *Massetognathus*.

The therapsids continued to flourish worldwide throughout
the Upper Permian. Then, 250 million years ago, the greatest
cataclysmic event in the history of life struck: the mass extinction
marking the end of the Permian. In the sea and on land over
90 per cent of the animals and plants disappeared, and the
therapsids suffered as much as any other taxon. The trigger for the
mass extinction was a period of huge volcanic eruptions. Volcanic
gases poured into the environment, causing catastrophic global
warming due to the greenhouse effect of the carbon dioxide, and

acid rain from sulphur dioxide killed off much of the plant life. The environmental conditions for a long time afterwards were hot and arid, and most of the lush forests of the earlier, benign times had disappeared.

Almost miraculously, a few therapsids managed to survive the catastrophe and we find fossils of them in rocks that date from the very start of the succeeding Triassic Period. One is a dicynodont called *Lystrosaurus*, a burrower living on underground storage stems and tubers, and presumably it was this habitat that saved it. During its brief existence, *Lystrosaurus* holds the record as the most widely distributed terrestrial tetrapod of all time, having been found in rocks of every one of the modern continents, even Antarctica. One or two other dicynodonts and three or four small therocephalians also survived, probably because they were burrowers as well.

Mammals

But, for our story, by far the most important therapsids to survive into the Triassic were one or two cynodonts. From a fairly insignificant role in the Permian fauna, the cynodonts evolved throughout the Triassic, and by the end of the Period, had given rise to the mammals.

Cynodonts

Thrinaxodon is the commonest cynodont fossil found immediately after the mass extinction. It was a small, active animal like *Procynosuchus* and even more mammal-like. Its molar teeth had much bigger extra cusps. The larger temporal fenestrae and the deeper sagittal crest between them show us that the jaw-closing muscles were stronger still, and the tooth-bearing dentary bone of the lower jaw had a large extension upwards (*coronoid process*) for them to attach to. By now, the jaw was being closed by a sling of muscles, as we saw earlier in modern mammals (Figure 4), made up of the temporalis muscle attaching to the inner side and the masseter muscle attaching to the outer side. The force and the accuracy of the bite were both increased. The lower jaw was made up of several bones: the dentary, as its name implies, carrying the teeth; and the bones behind the dentary, called the *postdentary bones*. The dentary bone of *Thrinaxodon* was enlarged in size and its postdentary bones correspondingly reduced. This includes two bones right at the back that make up the jaw hinge, the *articular* below and the *quadrate* above.

In several other features, *Thrinaxodon* was more mammal-like than *Procynosuchus*. For example it had a well-defined ribcage, showing that a larger lung and probably a diaphragm had evolved; longer, more slender limb bones; and a shorter tail.

The therapsids never again enjoyed the dominance of the land they had enjoyed in the Upper Permian, because by now another group had arrived on the scene, and was spreading and competing for the habitat. These were the early archosaurs that eventually gave rise to the dinosaurs and crocodiles. Nevertheless, the cynodonts did continue to evolve, and by the Middle Triassic had become an even more mammal-like group called the eucynodonts. *Chiniquodon* (Figure 9(b)) from South America is a good example. It was a sharp-toothed carnivore about the size of a wolf. It had a huge temporal fenestra for the jaw muscles, and these were now attached only to the dentary. This bone was by far the largest of the lower jaw bones and the rest, the postdentary bones, were reduced to a narrow rod set in a groove along the inner face of the dentary. By this stage, virtually all the force generated by the sling of jaw-closing musculature was concentrated at the point of the bite between the upper and lower molar teeth, with almost no pressure being exerted at the jaw hinge. The jaw muscles had finally achieved the combination of an extremely strong bite with the small, very precise sideways and to-and-fro movements of the jaw necessary for chewing teeth to work. The molar teeth of *Chiniquodon* were adapted for slicing, with a sharp rear edge and one or two additional cusps at the back of the blade. Other eucynodonts had broad molar teeth, each one with a wide crushing basin bounded by a sharp cutting edge, suitable for masticating plant material.

We see other important advances towards the mammals in the skeleton (Figure 9(c)). The upper bone of the hip girdle, the ilium, was expanded forwards while the lower part of the girdle and the tail were reduced. This is because most of the muscles were high up and in front of the limb itself, so the hind leg could take longer, faster strides. The shoulder girdle was narrow and very loosely attached to the rib cage, so that the front leg also had the longer stride and greater flexibility of movement characteristic of mammals. It would be extremely interesting to know the details of the anatomy of the cynodont brain, given how important this organ was in mammal evolution, but unfortunately it was still too small to leave a detailed impression on the inner bones of the skull. What we do know is that the overall size was no larger than that of a reptile and therefore the expansion of the neocortex had yet to happen. It did, however, have an enlarged, mammal-sized hind brain, the cerebellum. This is the region responsible for fine control of muscle actions, so it is not surprising that it was the first part of the brain to enlarge, in view of the evidence we have seen for precise movements of the jaws, and speed and agility of locomotion in cynodonts.

We see the final stage before the full transition to mammals in the Upper Triassic fossils of *Brasilitherium* and several other extremely mammal-like cynodonts discovered in South America. They were very small, rat-sized animals whose most important new mammalian characteristic was a contact between the dentary bone and the cheek region of the skull. This came about by continued expansion backwards of the dentary that started in the earlier cynodonts, until it reached the squamosal bone in the cheek. The contact is the beginning of the new mammalian jaw hinge, and the small original jaw hinge between the quadrate and the articular still existed alongside it. Another notable mammalian characteristic of *Brasilitherium* is the loss of the bar of bone behind the eye socket.

The earliest mammal

One of the most exciting of all fossil finds was made half a century ago, in 200 million year old deposits in a South Wales quarry. It consisted of many hundreds of fragments of the teeth, jaws, and bones of a tiny, mouse-sized animal called *Morganucodon*, which were about forty million years older than almost any other fossil mammal then known. Since then, specimens of other, related morganucodontids, including complete skeletons (Figure 10(d)), have been found in Africa, China, and other places.

By this final stage in the evolution of mammals, the simple contact between the dentary and squamosal bones we saw in *Brasilitherium* had become a new, strong, ball-and-socket hinge. The original hinge bones, articular and quadrate, were tiny and easily recognized as the sound conducting ear ossicles, malleus, and incus, of later mammals, even though they still lay alongside the new joint. The morganucodontid dentition was completely mammalian in nature, with double-rooted molar teeth bearing three cusps in a row connected by sharp crests. As the jaw closed, the lower and upper molars come into precise occlusion, and the crests of the lowers worked against the crests of the uppers like the blades of miniature pairs of scissors. Another important new characteristic of morganucodontids was a fourfold increase in the size of the brain compared to cynodonts, due to an enlarged forebrain. By now the evolution of the neocortex was well under way. The morganucodontid skeleton is just like a typical unspecialized mammal apart from a few minor characteristics still left over from the therapsids, such as free ribs in the neck region and a simpler shoulder girdle.

As to the biology of morganucodontids, the small body size and sharp-cusped teeth are typical of an insectivore, and there are good reasons leading us to believe they were nocturnal. First and most simply, this is the mode of life of most small insectivorous mammals. Second, the mammal-like teeth, brain, and limbs all point to a mammal-like high metabolic rate as an adaptation for remaining active during the cooler night hours. Third, the most important senses appear to have been the two most useful at night: olfaction, as shown by the large olfactory bulb of the brain; and hearing, as evident from the evolution of the sensitive ear ossicles.

Our story of the origin of mammals, as spectacularly revealed by the fossil record, is about the evolution of many characteristics associated with an ever-increasingly energetic lifestyle. The culmination was a new kind of animal, one that could carry on being active at night as well as during the day, and in cool as well as warm seasons. Its newfound efficiency of hunting and feeding, and its enhanced awareness of its environment made sure that it could acquire all the extra nutrition it needed. And it had the potential to adapt these new skills by evolution to suit many new habitats: from this beginning the great mammalian radiation of the next 200 million years commenced.

Chapter 4
The radiation of mammals

The Mesozoic mammals

With the benefit of hindsight, we know how enormous was the potential of this little *Morganucodon*-like ancestral mammal of 200 million years ago to give rise to the huge diversity of forms and lifestyles of its descendants, not least ourselves. What might still surprise us though is just how long a time it was before the great range of mammals we see today actually appeared. For no less than two-thirds of their entire history, right through the 135 million years left of the Mesozoic Era (that is, the Jurassic and Cretaceous Periods), mammals were all small, secretive animals feeding at night on insects, worms, seeds, and plant tubers. Most were the size of mice, rats, and rabbits, with just one or two as big as a domestic cat. The predatory mammal *Repenomamus giganticus* was the one unique 'giant' amongst them, but even its body excluding the tail was only 60 cm long, and its weight 12–14 kg, similar to a European badger. The problem the mammals faced during the Mesozoic was that they co-existed with the dinosaurs, whose own radiation had started at about the same time. Dinosaurs happened to be the first animals to evolve into middle- and large-sized land animals, and due to what is called *competitive exclusion*, they prevented mammals from evolving this way of life. On the other hand, mammals were the first to adapt to the high-energy, nocturnal way of life of a small animal, and so by the same process they prevented dinosaurs from evolving very small body sizes.

We would, however, be mistaken if we thought that mammals did not evolve very much during the Mesozoic because of the restriction on their body size. As more and more Mesozoic mammals have been discovered, especially the beautifully preserved, whole skeletons, complete with an impression of the fur, from deposits in China (Figure 10(a)), it has grown clear that the Mesozoic forests and riversides teemed with small mammals, just as they do today. Many new kinds of teeth and limbs associated with different habitats evolved. Most groups had sharp molar teeth for chewing up small invertebrate prey. However, the most abundant and long-lived of all were the rodent-like multituberculates (Figure 10(b) and (c)), which had huge incisors at the front for collecting food, followed by a massive blade-like premolar for chopping it, and broad molars and premolars bearing rows of blunt cusps for finely grinding it up. They lived on a diet that included energy rich plant food, such as seeds and tubers. Another group, the eutricondonts including the relative giant *Repenomamus*, evolved triple-cusped cutting postcanine teeth for dealing with their small animal prey, as the fossil of the largest specimen dramatically proves. It is preserved to this day with its last meal still within its body cavity—a baby dinosaur.

In addition to the variety of feeding adaptations, mammals with different means of locomotion evolved. There were diggers with short, powerfully clawed limbs, and jumpers with elongated hindlimbs. One was a small beaver-like animal appropriately called *Castorocauda* because of the large flattened tail for swimming, while several were arboreal animals with grasping feet and a prehensile tail. *Volaticotherium* even had a gliding membrane like that of a modern flying squirrel.

The Chinese deposits have also yielded the earliest fossils of what were destined to become the two main kinds of living mammal. The 135 million year old *Eomaia* (Figure 10(a)) is a placental, and its contemporary *Sinodelphys* a marsupial. Although looking quite similar to one another, these two already differed in the numbers of each kind of tooth and the arrangement of the cusps on the molar teeth that separate the modern groups.

Then everything changed forever as the next great cataclysmic event hit the world. Sixty-five million years ago a mass extinction

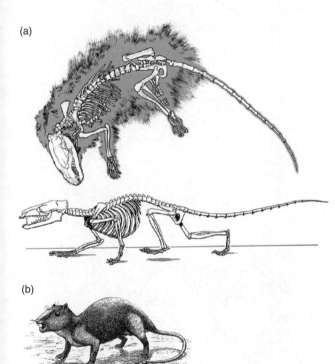

(a)

(b)

10. Mesozoic mammals: (a) *Eomaia*, the earliest placental mammal as preserved with pelt impression, and restored; (b) reconstruction of the multituberculate *Nemegtbaatar*;

(c)

(d)

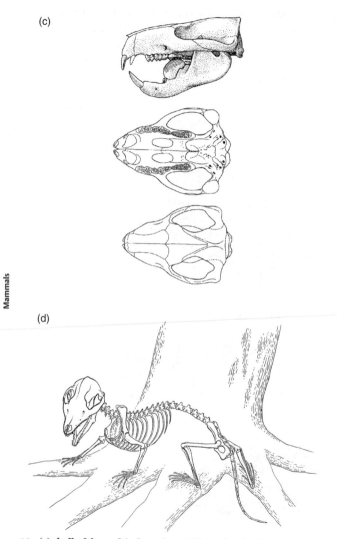

10. (c) skull of the multituberculate *Ptilodus* showing the specialized, rodent-like dentition and skull shape; and (d) reconstructed skeleton of the morganucodontid *Megazostrodon*.

marked the end of the Cretaceous Period, and with it the close of the Mesozoic Era. Over 60 per cent of all the living species disappeared, and the dinosaurs (apart from the birds, which are technically a branch of dinosaurs) were obliterated, along with the pterosaurs in the air, and the ichthyosaurs and plesiosaurs in the seas. The cause of the mass extinction is still energetically debated amongst palaeontologists. Some are content to believe that it was a massive meteor that crashed into the sea just off the Mexican coast; others point to massive volcanic activity at the time as being the prime culprit. Whatever it was, the fact is that out of the rich land vertebrate fauna at the end of the Mesozoic, only a few multituberculate, placental, and marsupial mammals survived, along with a handful of reptiles, birds, and amphibians. But this was enough to seed the next phase of mammalian evolution.

The Tertiary radiation of mammals

Scarcely had the Earth recovered from the end-Cretaceous extinction than the new mammalian radiation was underway. Within no more than two to three million years after the start of the Tertiary, mammals characterized for the first time by a larger body size began to appear in the fossil record. In a world recently cleared of dinosaurs, there was no longer any competition to prevent mammals from evolving into large, active animals that could spend the daytime feeding on leafy plants. Several early placentals called condylarths developed larger, blunter molar teeth for this new diet, and small hoofs replaced sharp claws at the ends of their toes to improve their running (Figure 11(a)). Most were around the size of a rabbit, but some evolved into larger animals, such as *Phenacodus*, which at 50 kg was as big as a sheep. Members of some early placental groups reached over 100 kg in weight, such as the omnivorous taeniodonts and the more specialized, herbivorous pantodonts (Figure 11(b)). Other placentals soon evolved into larger carnivores to prey on these new herbivores. Creodonts (Figure 11(c)) grew to the size of a wolf,

11. Larger bodied mammals in the Paleocene: (a) the condylarth *Meniscotherium*; (b) the pantodont *Coryphodon*; (c) the creodont *Machaeroides*.

and had sharp claws, powerful canine teeth, and carnassial molars with strong slicing crests for dealing with their prey.

Several other new groups evolved during these early Tertiary times, but they remained small in size. These included the first members of some of the modern placental Orders, Carnivora, Primates, and Rodentia; and the insectivorous group Eulipotyphla that today includes the shrews, moles, and hedgehogs. However, it was the Early Eocene epoch, about fifty million years ago, that witnessed the greatest diversification of mammals. It was a time of high global temperatures, around

13°C higher than those of today. This warm climate, coupled with the high level of carbon dioxide that mainly caused it, encouraged a huge increase in plant life. Rich forests covered great swathes of the Earth's surface, even in the high latitudes of Antarctica to the south and Alaska to the north, and these were home to an abundant fauna. Mammalian diversity has never been so great, either before or since. Groups that had survived from the Mesozoic, the small omnivorous multituberculates, the opossum-like marsupials, and the primitive insectivorous placentals, continued to flourish, whilst the new, early Tertiary groups expanded alongside them. Almost all the remaining placental Orders make their appearance in the fossil record of this time. The five-toed, terrier-sized horse *Hyracotherium* ('*Eohippus*') represented the perissodactyls, while the similar-sized antelope-like *Diacodexis*, the artiodactyls. There were small, cat-like Carnivora called miacids, and the very earliest elephant, *Phosphatherium*, that was a mere 30 cm high, lived in North Africa.

The most specialized of all mammalian Orders also evolved in the Eocene. We should not be surprised, except perhaps at the good fortune of discovering them, that several fossil whales from coastal deposits of what is now Pakistan still had substantial limbs. *Pakicetus* (Figure 12(a)) could still run fast on land. *Rodhocetus* (Figure 12(c)) had long, paddling limbs and, like modern seals, was clumsy on land but an adept swimmer. Meanwhile, a long legged, 2-metre sirenian called *Pezosiren* (Figure 13) was wading around the shores of the West Indies, and the earliest discovered bats had already perfected their flight mechanism, although they were not yet capable of echo-location.

Towards the end of the Eocene the world's temperatures and rainfall began to fall, and the climate became altogether less favourable for mammals. About thirty-two million years ago they suffered a wave of extinction called '*La Grande Coupure*', when over half the species disappeared. The older mammalian groups

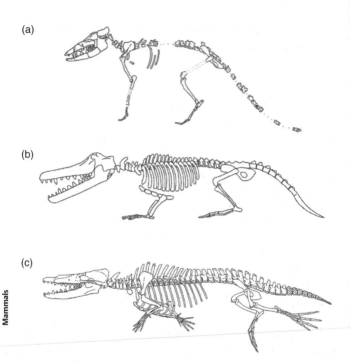

12. Eocene limbed whales: (a) *Pakicetus*; (b) *Ambulocetus*;
(c) *Rodhocetus*.

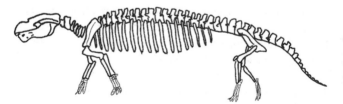

13. The Eocene sirenian *Pezosiren*.

suffered most, as the last of the long-surviving multituberculates disappeared along with nearly all the condylarths and other early kinds of herbivores, and the creodonts.

The crisis ushered in the Oligocene epoch, during which few major new mammalian groups evolved. But the start of the Miocene twenty-three million years ago was marked by a return to warmer, more favourable conditions. A rapid radiation of the mammalian groups that had survived followed, and before long their diversity reached the second highest level in mammalian history. Although it was the modern Orders with which we are familiar today that populated the world, there were some decidedly unfamiliar members. For example, the hyraxes are a very insignificant Order of small mammals now, but during the Miocene they were one of the most important groups of large herbivores, and included some as large as rhinoceroses. Amongst the perissodactyls were horses, tapirs, and rhinos that would not look out of place in the modern world, but also some of the most bizarre mammals of all. *Paraceratherium* (Figure 14) measured 6 metres at the shoulder and weighed 40 tonnes, making it the largest mammal ever to have lived. *Chalicotherium* was an equally strange giant perissodactyl. Its front legs were almost twice as long as its hind legs, and it carried huge, grasping claws to pull down branches and feed on the leaves.

Grasses are a nutritious, but tough form of plant food. During the Miocene, they expanded over great areas, creating the North America prairies, South American pampas, Eurasian steppes, and African savannahs. Mammals can make a very good living from grass providing they have large, high-crowned teeth, resistant to excessive wear. Miocene horses such as *Miohippus* had such hypsodont teeth, but by far the largest number of mammals possessing them were artiodactyls (deer, antelopes, and so on),

14. The Miocene perissodactyl *Paratherium*, the largest known land mammal.

and many new species evolved to take advantage of this newly abundant diet.

Amongst the Miocene carnivorans preying on these herbivores were several different sabre-tooths, with their huge canine teeth for killing large-sized prey (Figure 15).

Another radically new version of placental first appeared during the Miocene. With a reduced snout, enlarged brain, and elongated forelimbs, primates such as *Proconsul* were the earliest of the apes, and represented the start of the hominoid radiation that was eventually to include humans.

The island continents—Australia and South America

During the Tertiary there was plenty of opportunity for mammals to disperse widely between the northern land masses of North America, Europe, and Asia, and by the Miocene Africa had drifted northwards to join them. The main Orders of boreoeutherian

15. The Miocene sabre-toothed *Barbourofelis*.

mammals became widely distributed throughout, while the afrotherian elephants and a few hyraxes spread northwards. However, our picture of mammalian evolution is still incomplete, because there were two large, geographically isolated island continents: South America, until a mere three million years ago when it joined North America, and Australia. Mammals in each of these evolved in isolation from the rest of the world, resulting in quite different groups.

In the case of Australia, by far the majority of mammals were and still are marsupials. The only placentals to have arrived before humans introduced dingos, rabbits, and others, were bats, which could fly there, and a number of rodents, which got there in the last few million years by chance from Southeast Asia. The marsupials evolved from a group of opossum-like ancestors that occupied the great southern supercontinent called Pangaea. They remained isolated in Australia when it broke away from Pangaea, and by the time of the superb twenty-million-year-old Miocene fossil record of Riversleigh in Queensland, they had evolved into many different kinds. Small insectivorous opossums, rabbit-sized bandicoots, larger herbivores such as wombats and kangaroos, carnivores such as the thylacines, and even a marsupial mole all made up a fauna very similar to the one we see in Australia today.

The evolution of mammals in South America is a good deal more complicated and interesting because this continent was not quite as isolated as Australia, and because we have a much better sequence of fossils. The two Americas were connected until about the time of the end-Cretaceous mass extinction, at which point South America broke away and started to drift southwards. One group of small North American didelphid marsupials was already present, and over the next tens of millions of years these evolved into the South American opossums. They also gave rise to a group of marsupial carnivores called borhyaenids that ranged from fox-like animals to others as large as bears, and there was even a large, predatory sabre-tooth called *Thylacosmilus*.

Unlike in Australia, placental mammals were also living in South America by the start of the Tertiary. There was a variety of medium- to large-sized herbivores that had evolved a grinding dentition of enlarged cheek teeth, and hoofs in place of sharp claws on their toes. Some, the litopterns, resembled horses, even to the extent that later members evolved high-crowned teeth for dealing with grasses, and a single toe. Others such as *Pyrotherium* were enormous animals resembling elephants, complete with long tusks. The notoungulates were the most diverse, from rabbit- and sheep-like to the huge, bear-like *Toxodon* (Figure 16(b)), which had the size and proportions of a large rhinoceros.

The Xenarthra, one of the modern placental Superorders (see Chapter 1), is another group of uniquely South American placentals. The living armadillos, sloths, and anteaters are strange enough animals, but they are nowhere near as odd as some of the

16. South American Tertiary mammals: (a) the giant ground sloth *Megatherium*; (b) the notoungulate *Toxodon*; (c) the xenarthran *Glyptodon*.

extinct members. The 2-metre-long glyptodonts were relatives of armadillos that looked like tanks in their massive armour-plating (Figure 16(c)), while *Megatherium* (Figure 16(a)) was a giant ground sloth as big as an elephant, and could reach foliage 4 metres above the ground by standing on its hind legs and pulling it down with its huge front claws.

In rocks about thirty to thirty-five million years old, two surprising new groups of fossils suddenly turn up, first rodents and, soon afterwards, primates. The only place these could possibly have come from is Africa: they must have been carried on islands of floating vegetation torn from the shore by a storm and, by good luck, drifted all the way across the southern Atlantic Ocean. Once they got to their new home, they rapidly evolved into an important part of the fauna that we still see today. The South American rodents (Hystricognatha) include various kinds of rats, porcupines, and the water-loving capybaras, one of which, the extinct *Phoberomys*, was a hippo-sized giant weighing 700 kg. It was by far the largest rodent ever to have lived. Other hystricognaths, such as the mara, evolved longer legs and fast running, just like small antelopes. Meanwhile, the primates diversified into the New World branch of monkeys, the broad-nosed Platyrhini, such as spider monkeys, tamarins, marmosets, and the howler monkey.

The final phase of the story of South American mammals began about three million years ago. After drifting northwards for some time, the continent finally reconnected with North America, and many mammals dispersed in one direction or the other across the Isthmus of Panama. The biggest shift was of northern mammals entering South America, where elephants, carnivorans, deer, camels, horses, and others promptly established themselves, and played a part in the extinction of most of the southern groups. Of the original South American mammals, only some opossums, the smaller xenarthrans, and the New World rodents and monkeys still survive. Indeed, a good few species of these mammals

successfully spread northwards, where they are now part of the
modern mammalian fauna of Central America and southern USA.

The endgame

Average worldwide temperature and rainfall have been slowly
falling over the past five million years, accompanied by a gradual
decline in the number of mammals. On top of this, mammals
suffered a more sudden and extreme phase of global extinction
that started about 60,000 years ago—coinciding with the end of
the series of Pleistocene ice ages. It was not a particularly large
catastrophe in terms of number of species lost, compared to some of
the geologically earlier times, but the *end-Pleistocene* extinction was
different because it was very strongly biased against large bodied
mammals: most of what we term the *mammalian megafauna* was
lost. The event coincides with changes in climate and vegetation
as the great ice sheets retreated, and also with the approximate
timing of the spread of humans around the globe. Whether the
megafaunal extinction was due to natural, environmental change
or to over-hunting by humans, even at this early period of their
expansion, has been argued over for more than 150 years. We
shall return to the question in the last chapter, but meanwhile,
whatever the cause was, the fact remains that it was the final
important factor shaping the mammalian fauna we have today
that we shall examine in the next few chapters.

Chapter 5
Carnivorous mammals

Insectivorous mammals

We think that the common ancestor of all the modern mammals was small and mainly active at night, and judging by their teeth, fed on insects, worms, and other invertebrates. Several kinds of mammals still follow this mode of life today, such as the placental shrews, moles, hedgehogs, the tenrecs of Madagascar, and, amongst the marsupials, many of the opossums of South America and Australia. All have sharp, pointed incisors and canines for capturing the prey, followed by sharp-crested premolar and molar teeth. Once the food is in the mouth, the first action of the jaws is a simple mashing action, using the premolars and molars. If the food is tough, such as insect cuticle, a second stage follows consisting of a more precise cutting action between the upper and lower molar teeth, whose opposing sharp crests act like scissor blades and reduce the food to small particles.

Insects and other small animal prey are highly nutritious and readily digested by protein-dissolving enzymes so the intestine can be short and simple. The main problem with a diet of insects is capturing them in the first place. The important attributes to achieve this are acute sensitivity to their presence, particularly their sounds and smell, and sufficient agility to capture them. Small body size helps, because the small size of each individual insect makes it difficult to acquire enough to satisfy the needs of a larger animal (although ants and termites can be an exception, as we shall see in a later chapter).

Larger carnivorous mammals

The evolution of a small, insectivorous ancestor into larger bodied, predaceous mammals required relatively few anatomical changes to the dentition. The main modification is that the V-shaped cutting ridges of one or more of the back teeth tended to become exaggerated and aligned longitudinally, creating what are termed *carnassials* (Figure 17(a)–(c)). These act as blades for slicing through the prey's tough tendons holding its muscle mass together, and can cut and tear up the meat into easily swallowed pieces. Less highly specialized carnivorans such as dogs (Figure 17(a)) still do have a broader part to the back molar teeth capable of a crushing action, because they have a more mixed diet that includes berries, invertebrates, and carrion as well as freshly killed meat. Hyaenas too (Figure 17(c)) have broader molars, although in their case it is for crushing bones, with the help of massive jaw muscles, to get at the marrow. Cats (Figure 17(b)) are the most extreme carnivorans of all, and the teeth behind the large canines are reduced to little more than one or two pairs of large carnassials, indicating an exclusively carnivorous diet. In the opposite direction, some carnivorans such as badgers tend towards a diet containing more plant material, consuming roots and tubers along with earthworms. The bears too (Figure 17(d)) have adopted a truly omnivorous diet. Their carnassial teeth have become broad and blunt, in keeping with a diet that includes large amounts of berries, roots, tubers, and often fish, in addition to meat. The carnivore dentition has been modified in yet another direction by the aardwolf, which has only a few simple teeth. It is a small hyaena of the African grasslands (Figure 18) living on ground-dwelling termites. These it detects using large, extremely sensitive ears, and licks them up with a long, sticky tongue.

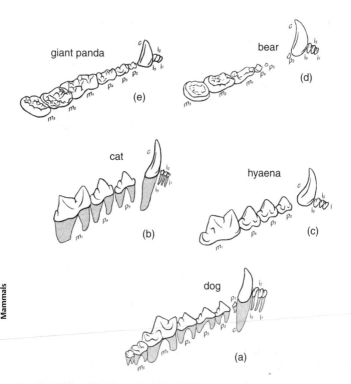

17. Variation of the lower dentition of Carnivora: (a) dog (*Canis*); (b) cat (*Felis*); (c) *Hyaena*; (d) bear (*Ursus*); and (e) giant panda (*Ailuropoda*).

A large predaceous mammal needs no great changes to its intestines from that of its insectivorous ancestor, and the main adaptations are for capture of its prey, which may often be of a comparable or even larger size than itself. Keen eyesight is essential, and the eyes must face forwards for accurate *binocular* judgement of distance during the chase and attack. The two broad solutions for capture of large prey are stealth hunting by solitary animals, or pack hunting by organized social groups.

18. Aardwolf foraging.

Most of the members of the cat family, Felidae, are solitary hunters like tigers and wild cats, although lions are something of an exception. They do hunt mostly as a group, but solitary individuals are perfectly able to fend for themselves, and even within a pride a single female can bring down an adult buffalo on her own before the rest of the group move in to help her complete the process. A felid's limbs are powerful, muscular, and relatively short, which gives the animal rapid acceleration, although they are not very good for sustaining fast running. This compares with the long slender legs of the antelopes and deer that make up most of the large cat's prey. A typical felid therefore hunts under the cover of darkness or vegetation by first approaching its prey as closely as possible before detection, and only then attacking. The limbs need to have sharp claws for grasping and bringing down the prey, an anatomical feature that further reduces the animal's efficiency of running. If the initial attack fails, further pursuit can only be of short duration as the predator will rapidly tire. While this is the strategy of hunting for the majority of cats, the cheetah has evolved longer, more lightly built legs making it capable of higher speeds, though still in relatively short bursts. They can reach speeds of up to 100 km per hour and are able to hunt successfully in open plains and grasslands where they are more visible, but ultimately they too are limited by a lack of stamina.

The dog family, Canidae, illustrates the alternative style of predation: organized, communal hunting. African painted dogs, for example, live in packs of ten to twenty-five individuals, and social coherence amongst the individuals is maintained by elaborate tactile and vocal greeting rituals. There is a dominant alpha pair, male and female, who direct the communal hunting operations to a surprising degree of detail. The prey can be animals such as buffalo or zebra, which are much larger than the dogs, although more typically it is smaller antelopes. Most of the pack will set off in pursuit of a selected prey, but if for example the quarry turns away to one side, some of the dogs from the back will peel off in a coordinated way to intercept it further ahead. With their relatively long, slender legs, painted dogs can maintain a speed of 50–60 km per hour several kilometres as they chase an antelope, before one of the pack gets near enough to grab it in the rear and the rest can then assist in bringing it down. In the northern hemisphere, wolves exhibit the same degree of social cooperation in food capture.

Chapter 6
Herbivorous mammals

Compared to a predator's diet, plant food has two great advantages: it is abundant and it does not run away. So we should not be surprised at how many mammal species are herbivores, taking advantage of the huge variety as well as abundance of vegetation. These advantages are, however, matched by several difficulties. First, plants are generally of low nutritional value compared to meat and must be eaten in large amounts, especially when it is foliage. Second, leaves often contain protective abrasive particles, and if they grow near the ground they are likely to be sandy or gritty, so they quickly wear down the herbivore's chewing teeth. Third, plant cell walls are made of the carbohydrate cellulose, and mammals, like almost all animals, cannot make their own cellulase enzyme for breaking it down to sugars. Instead they must rely indirectly on the activity of microorganisms in their gut for digesting it and releasing the cell contents. Therefore, to be a successful herbivore, a mammal must evolve special adaptations for ingesting and masticating the food in bulk, and for housing a population of bacteria in a part of its gut called a *fermentation chamber*, to digest the cellulose. Furthermore, herbivores need adequate protection against predators during feeding, when they are often particularly vulnerable to attack.

Small herbivores: rodents, rabbits, and hyraxes

Throughout their history, small bodied mammals adapted for eating the particularly nutritious parts of plants such as seeds, fruits, and storage organs have existed, for example the Mesozoic multituberculates we met in an earlier chapter. Today this way of life is followed most exuberantly by the rodents, often as part of a mixed diet that also includes invertebrates, possibly small lizards, and birds' eggs, and in the case of the extremely successful rats, almost anything they can get hold of. Rodents are in fact the most diverse of all mammals, representing no fewer than about 2,000 of the 5,500 living species.

In large part they owe their success to their modified dentition and jaw muscles (Figure 19(a) and (b)). The first pair of upper and lower incisors are long chisels whose gnawing action can cut up the toughest of foodstuffs, as well as making a very versatile tool for picking up items of food. The front edge of each incisor is covered in a thin layer of an extremely hard material, enamel, which always stays as a sharp cutting edge however much the teeth become worn with use. The roots of the incisors are set deeply into the jaws and the crowns grow continuously outwards to maintain their length, again despite all the wear resulting from the tooth to tooth contact. There is a gap in the tooth row, called a *diastema*, in place of the remaining incisors and canines, behind which the premolars and molars have broad, flat, grinding surfaces marked by a pattern of enamel ridges (Figure 19(b)). The food is ground up between the lower and upper tooth rows as the huge jaw muscles (Figure 19(a)) pull the lower jaw powerfully forwards and backwards, an arrangement capable of dealing with virtually any kind of food, even hard wood or the tough outer case of a nut.

The digestive system of rodents (Figure 19(c)) includes a large *caecum*, which is a blind extension attached towards the back of the intestine. It is filled with microorganisms that break down the

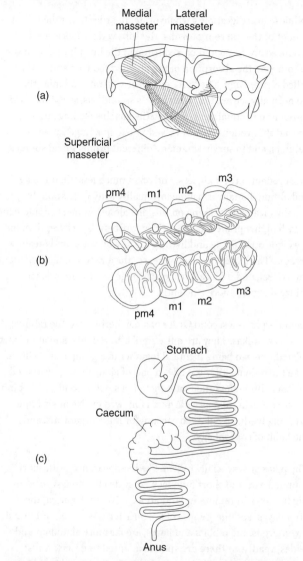

19. Rodent adaptations for feeding: (a) American porcupine showing the division and forwards disposition of the huge masseter muscle; (b) upper and lower left cheek teeth viewed obliquely; (c) rodent alimentary canal.

cellulose into digestible carbohydrate. In order to get the fullest benefit of the microorganisms, the soft, watery faeces first produced are eaten and passed through the intestines once more to complete the absorption of the digested food, a habit that is called *coprophagy*. The hard, dark faeces of rats and mice that we normally see are the result of this second passage of the food through the animal's gut. So versatile has this feeding strategy proved that rodents have evolved to live in a huge range of habitats, and to survive on many different primary food sources.

Most rodents are small, less, and often much less, than 1–2 kg, and although many consume low quality leafy vegetation, they are usually incapable of living on this diet alone but must supplement it with higher grade foodstuffs. The capybara is an exception, for it weighs up to 60 kg and lives by grazing on water and meadow grasses. The large body size increases the amount of food that it can collect, and it has a particularly huge caecum in order to utilize the grass more efficiently.

Lagomorpha is a second Order of small herbivores: the rabbits, hares, and pikas. They are quite similar to rodents, and are in fact related, the two being included together in a group called Glires. Like the rodents, they too have a pair of elongated, open-rooted incisors although in their case there is a small second pair behind the first. Again there is a diastema followed by five or six large grinding teeth, an enlarged caecum for microorganisms, and the habit of coprophagy.

The pikas are especially rodent-like in appearance, with short rounded ears and short hind legs. They dwell amongst rocks or dig burrows in remote upland areas of North America, the Himalayas, and the steppes of northeastern Asia. The rest of the lagomorphs, the rabbits and hares, are far more abundant and widespread, and there are species adapted to desert, forest, grassland, montane, and tundra habitats. As general herbivores surviving on a wide range of vegetation, they play an extremely important role in their communities, particularly in the more challenging habitats. They are the main primary consumers of plants, and the main source of food for carnivorous mammals and birds of prey. The legendary breeding of rabbits is no exaggeration, for they have several adaptations specifically to increase their reproductive rate and so compensate for the high rate of predation they suffer. Ovulation is stimulated directly by copulation instead of only occurring at fixed periods, and the female can conceive again immediately after giving birth, to reduce any delay in producing the next litter. The gestation period is only about a month, the litter size is usually large, and the offspring reach sexual maturity after a mere three months or so. Escape from predators relies on the extreme sensitivity of the long ears, coupled with high acceleration and fast running using the elongated hind legs: hares are able to reach speeds of up to 45 kph. This is important because most live in open ground, and only very few, such as the European rabbit, dig themselves burrows for safety and care of their young.

The Hyracoidea is the third Order of small herbivores. The dozen or so rabbit-sized hyraxes of today are a small remnant of a group that until about fifteen million years ago were the most abundant medium- to large-sized herbivores in Africa, occupying various roles similar to those of pigs, antelopes, and small rhinos. Indeed, they still have several features normally found in larger herbivores, such as high-crowned, hypsodont cheek teeth, a multi-chambered stomach for fermentation, and only four front and three hind toes, each ending in a small hoof. Hyracoids have rubbery pads on their feet to help them grip the smooth, rocky surfaces of their typical habitat, as for example the dassies (Figure 20) that are so familiar on Cape Town's Table Mountain. They survive on a poor diet of fibrous vegetation, helped by a low metabolic rate that reduces their nutrient requirements, and the production of highly concentrated urine minimizing how much water they need to drink.

20. Rock hyrax.

Large herbivores—'ungulates' and elephants

Two of the herbivorous Orders of placental mammals are loosely called 'ungulates' because they have hooves instead of claws on the tips of their toes, although they are not closely related to one another. Their premolars and molars are large and blunt, and form an effective chewing surface. Perissodactyla are the horses, rhinos, and tapirs; Artiodactyla are the pigs, deer, cattle, antelopes, camels, giraffes, hippos, peccaries, and chevrotains. There are good reasons why foliage-eating mammals tend to be large. They have lower metabolic rates compared to smaller animals and so need to eat relatively less food each day. And, thanks to their bulk, they are less liable to overheating in high daytime temperatures and therefore can continue to feed out in the open for longer periods.

The less specialized ungulates, such as the tapirs amongst the perissodactyls and the pigs amongst the artiodactyls, are more cosmopolitan in their diets, taking in all kinds of fairly soft vegetation and, in the case of the pigs, any invertebrates and carrion they may come across as well. Their cheek teeth are certainly enlarged, but the crowns have separate, rounded cusps and the roots are closed so that they do not grow continuously, limiting how much abrasive wear they can withstand.

About twenty million years ago, the great grasslands of the world appeared and offered a virtually inexhaustible food supply. Grasses, however, have evolved the protective device of embedding silica particles in their leaves making them extremely abrasive. Furthermore, being low-growing plants, they are often contaminated by sandy and gritty particles, adding to their abrasiveness. More advanced members of both the ungulate Orders evolved teeth capable of coping with this foodstuff, and these specialist grazers have become very successful and diverse as a result. The premolar teeth are similar to the molar teeth (Figure 21(a)), and each has a pattern of enamel ridges running through the full height of the crown and exposed on the large grinding surface (Figure 21(b)). As this surface wears away under the remorseless effect of chewing masses of leaves, the hard enamel ridges always remain slightly proud of the softer material of the rest of the tooth and therefore the tooth keeps its file-like roughness. The teeth are also very long and set deep into the jaws, and their roots are open and supplied with blood vessels. Each tooth continues to grow outwards as fast as its crown is worn away. The actual pattern of the enamel ridges differs from group to group, making it a useful characteristic for identification as well as showing us that hypsodont teeth evolved independently in different groups.

The jaw hinge of an ungulate is always flat and unrestricted so that the mandibles can move from side to side in the typical chewing action seen in a cow or horse, for example. The masseter or cheek muscle attaching to the outer side of the jaw is the largest jaw-closing muscle. It extends well forwards along the jaw to produce a very strong bite force, although restricting how widely the jaw can open. The soft cheeks covering it also extend well towards the front of the jaw, making it easier to retain the food in the mouth.

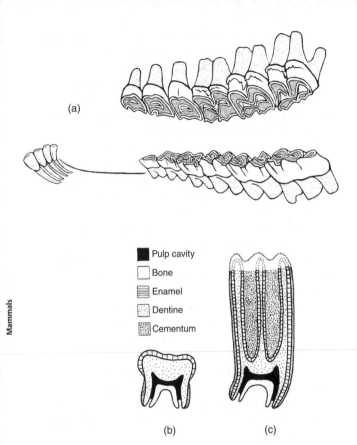

21. **Adaptations for herbivory in large mammals: (a) upper and lower dentition of a deer in oblique views to show the pattern of enamel ridges; (b) section though a low-crowned bunodont tooth such as a pig; (c) section through a high-crowned hypsodont tooth such as a horse.**

Several of the specialist grazers with hypsodont teeth also have modified incisors for collecting food faster. Horses' incisors are broad and blunt, and act like upper and lower gripping edges that can grasp a bunch of grass between them firmly enough for it to be torn off or pulled up. A number of artiodactyls use a different method, in which the upper incisors have been replaced by a horny pad. The lower incisors have sharp front edges that act as blades to cut against the pad, rather like the action of a knife on a chopping board, and so they actually crop the plants.

The mastication of the leafy food grinds it up and physically breaks down the cell walls of the plants to some extent, but most remain intact, and the cellulose it is made of is itself an important source of carbohydrate. Different parts of the gut in different ungulates are specialized as fermentation chambers, containing bacteria and other microorganisms capable of digesting the cellulose. The most diverse ungulates are the Ruminantia, the group of artiodactyls that includes deer, antelopes, cattle, sheep, and giraffes. As their name indicates, ruminants have the habit of regurgitating back into the mouth food that has already been swallowed, so they can chew it up a second time—'chewing the cud'. This is only possible because their fermentation chamber is a large, multi-chambered stomach at the front end of the gut (Figure 22(b)). The ingested vegetation is swallowed and passed into the stomach very quickly, with relatively little initial chewing. The animal later brings up this material from the stomach and grinds it up finely before it passes once again into the stomach chambers where it is exposed to the action of the microorganisms. Once these have done their work and the cellulose is properly digested, the food passes through the intestine to be absorbed. The advantage of this system arises because a herbivore faces constant danger of predation while it is feeding out in the open. Therefore the main work of mastication can be left until the animal is back in the safety of cover or amongst its herd.

However, using the stomach as a fermentation chamber in this way restricts the kinds of plants the animal can live on, because the stomach itself acts as a filter which only allows small, already digested particles to pass through into the intestine. The rate of movement of food through the gut as a whole is limited. Ruminants by and large must therefore eat relatively high quality food that

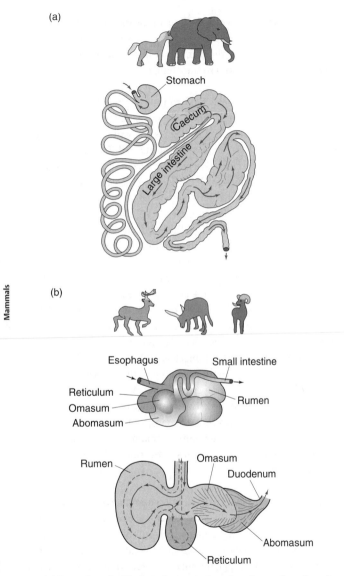

22. (a) Intestine of a hind gut fermentation chamber such as found in horses and elephants; (b) multi-chambered stomach of a ruminant such as found in antelopes, cattle, and sheep.

is low in fibrous material and can be broken down quickly into fine enough particles by the microorganisms. The alternative position for a fermentation chamber is the hind gut (Figure 22(a)), and this imposes no such limitation. Perissodactyls have a large, blind caecum attached to the large intestine towards the hind end of the gut, from which even quite large, incompletely digested food particles can be expelled. This incomplete digestion is less efficient because more food is wasted, but the disadvantage is balanced by the much greater total amount of food that can pass through the gut. Equids, the zebras, horses, and asses, can survive on tougher, lower quality fibrous plant material than can a ruminant because they can compensate for the low nutritional value by consuming a very high volume.

The elephants, Proboscidea, is the third modern Order of large herbivores. They are hindgut fermenters with a caecum (Figure 22(a)), and like the perissodactyls can survive on low quality vegetation. The elephant's most remarkable adaptation is its dentition. There are six huge cheek teeth and, depending on species and sex, each tooth consists of five to twenty-five enamel ridges, or *lamellae*, running across the surface of the crown. Instead of all the teeth functioning together as with most mammals, only one or two at a time are exposed in each jaw. As they wear down, they move forwards along the jaw until eventually the root is reabsorbed, and the remains of the crown are shed. Meanwhile, the next tooth behind has moved into place in the jaw and become the new functional one. By only using one or two enlarged teeth at a time on each side, the overall life of the grinding dentition is greatly increased, even though it is frequently called upon to deal with the toughest of vegetation. With the aid of its supremely versatile trunk, and the tusks which are a single pair of enlarged upper incisors, the elephant's diet includes the bark and twigs of small trees it may have uprooted or pushed over, as well as spiny acacia branches, grasses, softer foliage, and various fruits when in season. Such a catholic diet along with the large body size has enabled elephants to survive in a particularly wide range of habitats, including areas such as the Kalahari and the Namib Deserts, where at times food is both scarce and of poor quality. Yet they are equally at home in such lush regions as southern Asia and the rain forests of central Africa.

Small herbivorous mammals keep themselves as safe as they can by foraging over a small home range, and avoiding predators by inconspicuousness and often living in a burrow. Large herbivores require a large home range, and some, such as the wildebeest in Africa and caribou in North America, undertake huge annual journeys following the seasonal vegetation. They cannot hide from their predators and must rely instead on fast, efficient locomotion. How good this is depends on three things: the length of each stride, the number of strides taken per minute, and the efficiency of the conversion of the limb muscle contractions into the kinetic energy of locomotion.

The ungulate mammals evolved ways to improve all of these (Figure 23). The length of the stride is increased by simply elongating the limbs. It is also increased by moving the attachment of the limb muscles closer to the top of the limb, so that when they contract they swing the limb though a greater arc. A higher rate of striding is achieved by moving much of the weight of the limb, such as the muscle mass and the processes on the bones that it attaches to, high up near the top. This is like moving the weight attached to a clock pendulum higher up to make it swing faster, technically described as reducing its moment of inertia. This is also the reason why the lower, less weighty parts of the limbs are increased the most in length, and especially why the feet are extremely long, and held vertically, so that they make up about one-third of the limb's total length (Figure 23(a)). The ungulate therefore walks on the tips of its hoofed toes, which is called *unguligrady*. The weight of these feet is reduced by reducing the number of the toes. The perissodactyls are called the 'odd-toed ungulates' because the rhinoceroses and tapir have three toes with the third being the largest, while the horses only have the single third toe left. The artiodactyls are the 'even-toed ungulates', because their third and fourth toes are the equal largest and in most species they are the only toes present, forming the typical cloven-hoof.

The mechanical efficiency of ungulate running is increased because each time a foot hits the ground, some of the energy of the impact is not wasted but is used to stretch elastic ligaments between the joints (Figure 23(b)). These ligaments then contract, adding to the force of the next stride, just as a bouncing rubber ball stores energy as elastic deformation when it hits the ground, and yields it up as it rebounds. Another way of increasing efficiency found in all larger mammals, including humans but most extensively in ungulates, is having several different gaits, such as walking, trotting, and galloping, which are used for different speeds. Changing from one gait to another is a little like changing gear in a car. Muscles work most efficiently at one particular rate of contraction, and changing gait keeps the leg muscles operating as close as possible to this rate as the running speed changes.

Elephants have limbs described as *graviportal*, instead of the gracile unguligrade limbs of the ungulates. The reason for this is that the weight of an animal is related to the volume of its body, but the strength of the limbs that have to carry it is related to the area of their cross-section. It is easy to see from this simple mechanical principle how an animal that evolves an increase in its size must also evolve limbs that are disproportionately stronger and therefore thicker. By the time the 3–7 tonne weight of African elephants is reached, its limbs have had to become very stoutly built and be held completely vertically, and its feet must be short, strong, and make a broad contact with the ground.

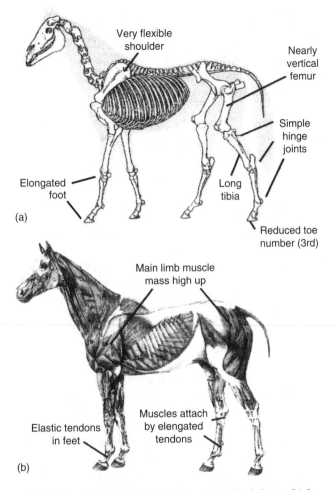

Very flexible shoulder

Nearly vertical femur

Simple hinge joints

Long tibia

Reduced toe number (3rd)

Elongated foot

(a)

Main limb muscle mass high up

Elastic tendons in feet

Muscles attach by elengated tendons

(b)

23. Adaptations for fast running in the horse: (a) the skeleton; (b) the limb musculature and tendons.

Marsupial herbivores: kangaroos, wombats, and koalas

The herbivorous marsupials of Australia are grouped together as the Diprotodonta, which includes the kangaroos, wombats, and koala, plus a variety of smaller species. They are so-named on account of the reduction of the lower incisors to a single pair of long teeth that are directed forwards and work against usually three pairs of vertical upper incisors. The 100 kg red kangaroo is the largest living diprotodont, and most closely matches the placental ungulates in adaptations for grazing on tough grasses and high fibre plant material. The lower incisors work against a horny pad in the roof of the mouth as well as against the upper incisors, creating a cropping mechanism. They are followed by a diastema between incisors and cheek teeth, and the latter are large with a square crown bearing enamel ridges, and growing continuously like the hypsodont teeth of ungulates. Furthermore, in a manner reminiscent of elephants, the cheek teeth gradually move forwards in the jaw. Front ones are discarded as they become worn out and new ones are added at the back. The kangaroo's stomach is a large, multi-chambered fermentation chamber, and like the ruminants, they can regurgitate their food and chew it up a second time before it passes into the intestine.

The speed of locomotion of kangaroos is increased, but in a very different way to the ungulates. Only the back legs are elongated, and the hopping is produced by rapid extension of both hind legs together. Elastic ligaments and tendons store as much as 50 per cent of the kinetic energy of each stride as the limbs flex on hitting the ground, which is then added to the energy of movement of the next stride. Kangaroos really do bounce along.

Wombats are small, bear-like marsupials weighing up to 40 kg. They feed mainly on grass, often of low quality, using their continuously growing cheek teeth and exceptionally powerful jaw muscles to grind it up finely. Unusually, the fermentation chamber is neither a stomach nor a caecum, as in other herbivores, but an extremely enlarged colon, which is part of the large intestine. They compensate for the poor nutrition they get from their food by having a low metabolic rate and a generally low level of activity, spending most of their time within the multi-entrance burrows which they dig with their powerful forelimbs.

Koalas also survive on a highly fibrous diet, in their case eucalyptus tree leaves, which, as well as having a low nutritional value, contain a number of substances toxic to other mammals. Uniquely, koalas are able to detoxify these in the liver. Fermentation of the food after it has been finely ground up by the cheek teeth takes place in a huge caecum. Like wombats, an important part of their adaptation for a poor diet is a very low metabolic rate, no more than half that of a more typical mammal of the same size. They are noted too for an extraordinarily low level of daily activity: a koala spends something like twenty hours a day sleeping in its tree.

Two strange herbivores—pandas and sirenians

The giant panda is a placental herbivore that, like wombats and koalas, exists on a particularly low quality diet. It is a member of the Carnivora and related to the bears, but is totally dependent on bamboo leaves for nutrition, one of the toughest of all foodstuffs. Its intake of shoots and leaves is helped by the celebrated 'Panda's thumb', which is not the sixth digit it looks like, but an elongated wrist bone. It works opposite to the rest of the hand for grasping in a similar way to the thumb of a primate. Even though a herbivore, the giant panda has the carnassial teeth typical of carnivorans for slicing up bamboo shoots and leaves, although the molar teeth behind are broad and flat for grinding (Figure 17(e)). Reflecting its fairly recent carnivorous ancestry, giant pandas do not have a fermentation chamber as such, but rely for cellulose breakdown on bacteria within the generally bear-like gut. This is a very inefficient system, and a giant panda must eat huge amounts. It spends as much as fourteen hours each day consuming 40 kg of bamboo, out of which only a fifth is actually digested. And it has to defecate up to forty times a day as the food passes through the gut so rapidly and in such bulk.

The most unexpected specialist herbivores of all are the dugongs and manatees. These, the Order Sirenia, are distant relatives of hyraxes and elephants, and like them they have a hindgut fermentation chamber in the form of a very large caecum. Sirenians are permanently marine, feeding principally on soft sea grasses, and we shall meet them again in Chapter 8.

Chapter 7
Diggers and burrowers

The ungulate mammals we met in the last chapter are specialized for high speed running. Their legs are long, slender, and held close to the vertical, and the muscles attach close to the top, near the joint between limb and limb girdle, so when they contract the stride is long and fast. We will now meet several mammals which are at the opposite extreme. They have short, powerfully built legs, the muscles attach well down the limb, and they bear large feet equipped with strong, blunt claws. When the muscles contract, they move the limb relatively slowly but powerfully, and so a large force is applied to the ground. Most of these animals are adapted for digging burrows to live in, and to find underground food such as beetle larvae or plant rhizomes. In a few, the limbs are adapted for digging into ant and termite nests, which can be extremely hard, so that particularly powerful muscles and claws are needed.

Burrowing mammals—several moles and the mole-rats

Many small mammals occupy underground burrows, where they gain protection from predators and extremes of ambient temperatures. Several amongst them have adopted a particularly specialized, more or less permanent version of the fossorial way of life, the most familiar being the moles, talpids, such as the common European mole (Figure 24(a)). The body is compact, the

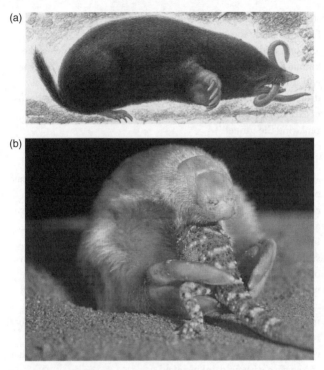

24. Moles: (a) European mole; (b) marsupial mole emerging from burrow and consuming a captured lizard.

tail short, and the tiny eyes are capable of detecting little more than the level of light around them. They have no external ears and their hearing is mainly by detecting ground-borne rather than air-borne sound. Their sense of touch, in contrast, is very well-developed, and the elongated, mobile snout is well-whiskered and highly sensitive to contact with prey. The mole's hands are huge, relative to its body size, and held out sideways instead of below the body. The short, robust bones of the forelimbs have strong muscles attached well down the shafts that move the hands powerfully from the forwards to the backwards position, shovelling soil behind in the process to allow the pointed snout to move forwards along the growing tunnel. The hindlimbs are smaller and press sideways against the sides of the tunnel to anchor the animal as it digs. Moles construct permanent tunnels in this way, and eat the various invertebrates, especially earthworms, that drop into them. The characteristic molehills they make are heaps of excess soil pushed up from time to time while digging or repairing the tunnel.

The golden moles (chrysochlorids) of southern Africa evolved similar burrowing adaptations to the true moles, a remarkable case of convergent evolution. They too have compact tailless bodies covered in dense fur, with short and powerful limbs, and small eyes and ears. Unlike the talpids, golden moles also have adaptations for living in highly arid regions where food and water are often scarce. Their metabolic rate and body temperature are low to save on food, and they enter a state of torpor for parts of most days, when their metabolic rate drops to an extremely low level, heart rate and breathing are reduced, and they remain completely inactive. This has the advantage of even further reducing their food requirements. Also, the golden mole's kidneys are so efficient at conserving water by producing highly concentrated urine that they can survive without drinking at all.

The marsupials of Australia include a single species of mole that lives in hot sandy deserts (Figure 24(b)). Once more there is a marked superficial resemblance to other moles, although the marsupial mole has several unique features, such as the fusion of the neck vertebrae into a single mass to strengthen the attachment of the head to the body for pushing through the sand. There are two huge claws on each hand, which, in addition to their digging function, are used to capture prey that includes small reptiles taken on the surface as well as underground insects.

Several kinds of rodent have taken up the mole-like habit, such as the aptly named African mole rats. Of these, the naked mole rats are the most notable, especially as far as their social behaviour is concerned, which is amongst the most remarkable of all mammals. In the shape of the body and the reduction of the eyes and ears they are fairly mole-like, but they are almost completely naked apart from a scattering of touch-sensitive hairs over the body. They have a very different mode of burrowing to that of moles too. Instead of using their limbs, they use a large pair of protruding incisor teeth for digging tunnels. Long, branched burrows leading to a breeding chamber are constructed by cooperation amongst a chain of individuals, the front naked mole rat doing the digging and the others disposing of the soil by using their feet to kick it backwards out of the entrance. They subsist entirely on vegetable matter, mainly underground tubers and roots that they come across while burrowing. The highly organized colonies are made up of several different castes, each caste having a different role. Only one female is reproductively active, and she is accompanied by a small number of reproductive males. All the other individuals are non-reproductive males and females. Some of these are smaller and make up the worker caste, whose role is to maintain the burrows, forage, and bring food to the rest. Larger individuals form a soldier caste that defends the colony at times of threat. Such a degree of division of labour as this is termed *eusociality*, and elsewhere in the animal kingdom it is only found in social insects like bees and termites.

Ant and termite eaters—aardvark, pangolins, and anteaters

Eating ants and termites is another mode of life that requires a very well-developed digging ability. Potentially these colonial insects make a very high quality and easily digested diet, but for a mammal to be capable of breaking into a termite hill baked to a concrete-like hardness in the tropical sun, or digging out a subterranean ant nest, requires extremely strong limbs and claws, and therefore it must be a fairly large animal. But because ants and termites are individually small items of food, a mammal that is large enough to get at them must also be adapted for collecting the very large number of them needed to satisfy its nutrient requirements. For this they have a very long, sticky, protrusible tongue. Three unrelated groups of placentals have independently evolved this combination of powerful digging ability and protrusible tongue. They are the aardvark of southern Africa, the pangolins of Africa and Asia, and the South American anteaters.

The aardvark, with its long, rabbit-like ears, is the least modified of the ant eating specialists (Figure 25(a)). It has short, robust limbs and four or five large, sharp-edged claws that can rapidly dig into the ground or a termite mound. The snout is long and tubular in section, and contains simple, peg-like cheek teeth. The tongue is its main feeding device. It can extend up to 30 cm from the mouth, and is sticky with saliva in order to collect large numbers of the insects at a time—as many as 50,000 in a single night. They are swallowed directly into the stomach, without chewing, where a special muscular region mashes them up. The single species of aardvark is entirely nocturnal and is the only one of the ant eating mammals that constructs a burrow—for protection during the day when it is not foraging and for rearing its young.

The pangolins and the South American anteaters were at one time believed to be closely related to one another, but the molecular evidence shows us that the anatomical similarities they share evolved independently for ant eating. These similarities include complete absence of teeth, and a tongue that can be extended as much as 50 cm from a very small mouth. The food, however, is treated differently once consumed. An anteater crushes it between the tongue and a pad in the roof of the mouth before swallowing it, whereas a pangolin has scarcely any jaw muscles at all but instead grinds up the insects in a thickened, muscular region of the stomach with the help of some stones it has swallowed. Both groups possess two or three huge claws on their forelimbs for digging into ant and termite nests, and these also act as very

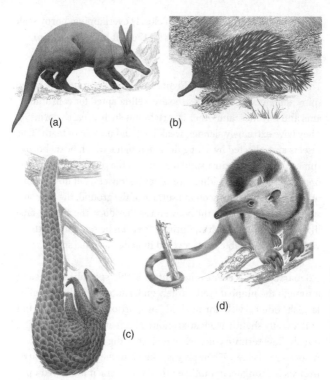

25. Ant eating mammals from four continents: (a) African aardvark; (b) Australian echidna; (c) Afro-Asian pangolin; (d) South American tamandua.

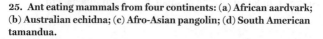

effective defensive weapons against predators. The tamanduas (Figure 25(d)), which are the smaller South American anteaters, and the pangolins (Figure 25(c)) are both very much at home in trees, with the help of a grasping, prehensile tail and the claws. This is particularly true of pangolins, which walk over the ground in a most ungainly fashion using the sides of their hind feet counterbalanced by the tail, to keep the sharp edges of their claws clear, but whose long slender tail is an impressive organ used for hanging from branches. The pangolin is also unique amongst living mammals in its covering of closely overlapping, horny scales protecting the whole body except for the soft belly and the inner surfaces of the limbs.

The echidnas of Australasia (Figure 21(b)) are appropriately called spiny anteaters, because they too are well-adapted for consuming ants and termites, and other invertebrates such as earthworms. They have extremely slender, weak jaws and no sign of teeth. The insects are collected by a long flexible tongue, which bears horny spines that work against similar spines in the roof of the mouth to break up the prey. Echidnas are adept burrowers, but do it in an unusual fashion. Sitting on the surface of the ground, the four short, stout limbs dig beneath the body and as they do so the animal sinks like a descending elevator, until all that can be seen of it is the virtually unassailable back covered in hedgehog-like spines.

No marsupial has ever evolved into a highly specialized anteater, although the numbat feeds largely on termites. It is the size of a largish rodent, with normal body proportions, a long bushy tail, and an only slightly elongated snout. Numbats forage by seeking out shallow termite colonies, which they dig into using their modest-sized claws. Their prey is taken into the mouth by a narrow, protrusible tongue, but unlike other ant eaters, it is chewed up using a complete dentition of small, rather simplified teeth.

Chapter 8
Aquatic mammals

Three hundred and sixty million years ago, an exciting new evolutionary venture began: a group of lung-bearing fishes began to swap their aquatic habitat for a life on land. Bit by bit, numerous adaptations developed, at first just enough for surviving short forays, but eventually, many millions of years later, mammals had evolved and were able to permanently withstand extremes of dryness, temperature, and gravity, and to thrive on dry land. Air had become their source of oxygen and carrier of sensory information, and water was no longer used for transferring sperm to egg cells. Stout legs for walking had replaced fins and tails for swimming, and endothermy protected them from the high daily and seasonal variations of air temperature. On the face of it, the multiplicity of modifications for living on land makes it inexplicable that so many mammals then returned partially or completely to an aquatic life. On reflection though, we should not be surprised at all. We have stressed all along that mammals owe their success to their relative independence of the environment, coupled with their extreme adaptability of behaviour. They were well-placed to take these two attributes with them into water, as another new environment to which they could adapt. Mammals did not evolve back into fish-like organisms but developed new ways of making use of the aquatic habitat. They never ceased to be highly energetic, endothermic, air-breathing, viviparous animals with large brains—quite unlike fish.

Virtually all mammals can swim at least sufficiently well to cross bodies of water, despite anecdotal claims to the contrary in this regard about camels, rhinos, and gorillas. The only probable exceptions are the giraffe, because its shape prevents it from adopting a stable, head-up orientation in water, and humans and apes, who have to actively learn the skill. The extent of specific adaptation for aquatic life varies amongst different mammals from no more than modified behaviour, like the Japanese macaque monkeys spending their days sitting in hot springs, through different degrees of amphibiousness to the permanently marine whales and dolphins, dugongs and manatees that cannot come out onto land at all.

Mammals of rivers and lakes

Mammals living in freshwater have evolved various degrees of aquatic adaptation. The least modified are certain rodents, such as the European water vole which has almost no specific adaptations for aquatic life other than a slightly flattened head, thicker fur, and a small amount of webbing between the toes of the hind feet. They live and rear their young in burrows they have dug low down on the river or lakeside bank, from where they forage for vegetation by swimming along the water margins.

The beavers are the most interesting semi-aquatic rodents because of the striking comparison between their unique dam building behaviour and our own tendency to modify the physical environment to suit our social needs. Another similarity beavers share with us is family structure. They live in small groups consisting of a life-long, monogamous breeding pair along with its recent young, and the offspring of the previous year which remain for two years and participate in the building activities, after which they leave to find mates and start their own families. Beavers are physically more modified for aquatic life than water voles, having a torpedo-shaped body, short front legs, fully webbed hind feet, and a flattened, scaly tail for swimming. Their eyes are protected by a transparent membrane, and they can close their ears, nostrils, and throat when gnawing beneath the surface. The fur is very dense and rich, offering waterproofing and insulation, and incidentally making it one of the most sought after animal skins by North American and European trappers. Their diet is strictly vegetarian, and, as well as eating succulent young plants in the Spring and Summer, they can survive on tree bark and wood, which they store over winter below the ice. Various engineering construction works are undertaken, including digging canals to connect adjacent bodies of water, and dams to maintain the water level in their living area. The actual living space is a lodge made of a pile of branches, logs, and finer vegetation built above water level, but which is safely entered by an underwater passage leading directly from the stream or lake.

The hippopotamuses in Africa, and the capybara (Figure 26(a)), which is a very large South American rodent, are other semi-aquatic mammals, but they use bodies of fresh water for protection from predators and for keeping cool during the day. Both are strict herbivores that come out onto land to feed. These mammals have slightly webbed feet which they use for swimming, but their ability in this respect is modest, and when in the water they prefer to simply sit on the bottom. Their nostrils, eyes, and ears are high up on the head so they can still use them even when almost completely submerged. The hippopotamuses are the more committed to water of the two, and they never feed more than 2 or 3 km away from it. They usually feed during the night and then spend most of the day keeping cool in the water, sleeping, and noisily communicating with other members of the often large, loose social group. They have lost their hair, and their skin is exceptionally thick and tough for physical protection; however, unlike the whales, hippos do not have a layer of insulating blubber. The way in which they do resemble whales is that the mother gives birth to, and suckles, her young under water. Capybaras are less

(a)

(b)

26. Freshwater mammals: (a) capybaras; (b) platypus.

tied to water than hippopotamuses. Their body is covered in thick, coarse hair, and they are often found in grassy meadows and forests. Although they normally mate in water, their young are always born on land.

The otters are more outwardly adapted for a semi-aquatic existence than any of the mammals we have seen so far. They are carnivores, belonging in the same group as stoats, weasels, and polecats, and share with them a long, slender body. This shape helps the land species to hunt for prey in burrows and confined spaces, but in the otters it streamlines them for moving through the water as they propel themselves with their webbed hind legs. The otter's tail is short and thick, and used for steering rather than for propulsion. Otters are small- to medium-sized mammals, and even the largest, the giant otter of South America, is little over 1 metre in length. This, along with the slender shape, makes them vulnerable to excessive heat loss in water, a problem that is counteracted as far as possible by a coat of dense hair including long guard hairs trapping a layer of insulating air. But they also have to have a high metabolic rate for adequate heat production, which in turn demands that they consume a lot of food. Consequently otters spend several hours a day seeking food. For most species, the main food is fish, particularly those that are more sluggish and bottom dwelling, like eels, which they detect using tactile hairs on the snout and capture with their teeth. Crabs and other invertebrates are also caught with the aid of very sensitive hands. The Asian short-clawed otter uses its fingers for grasping food items in a manner similar to the way humans use our opposable thumb and fingers. The sea otter of the North Pacific coastlands has the most remarkable feeding strategy of all, and is one of the few tool-using mammals other than primates. It can use a stone held between its paws to dislodge a bivalve mollusc, such as an abalone on the sea floor, by hitting it as often as necessary. Once it has freed the shell, the otter swims to the surface with it, carrying a suitable flat stone as well. Here it floats on its back with the stone on its front, grasps the shell between its paws, and bashes it against the stone until it breaks it.

Before we turn to marine mammals and their altogether more extreme adaptation for aquatic life, we should consider the oddest freshwater mammal of all, the duck-billed platypus (Figure 26(b)).

This member of the ancient monotreme lineage probably evolved its short powerful limbs for swimming from those of a burrowing ancestor, similar to its echidna relatives. The front feet are huge and fully webbed for providing propulsion, and the smaller, partially webbed hind feet are used for steering. The body is covered in an extremely dense coat of short, insulating hair. The foraging platypus dives to the bottom of a lake or river for two or three minutes at a time before returning to the surface to breathe. Despite its appearance, the bill-like extension of the snout is not at all like a bird's beak, but is soft and carries very sensitive tactile sense organs which, uniquely amongst mammals, can also detect the electrical impulses of its prey. It is only used for navigation and finding food under water. Once detected, worms, insect larvae, and snails are taken into the mouth, held in cheek pouches, and crushed between horny ridges on the jaws. Yet another curiously unique feature of the platypus is a hollow poison spur on the male's hind foot that connects to a venom gland and is capable of killing an enemy as large as a dog.

Platypuses dig tunnels as long as 30 metres into the bank, and in one of these the female lays two eggs, which soon hatch into juveniles. Here she feeds them with milk produced by her teatless mammary glands for two or three months until they are weaned.

Marine mammals—pinnipeds, sirenians, and whales

We can turn now to the height of aquatic adaptation in mammals, the three marine groups (a small number of which have invaded rivers systems, such as the Amazonian manatee and the Ganges dolphin). The pinnipeds are the sea lions, seals, and walruses, and are members of the Order Carnivora; the Cetacea are the whales, dolphins, and porpoises; and the Order Sirenia are the dugongs and manatees. All are relatively large, and most more or less hairless mammals streamlined for swimming, and of the three only the pinnipeds are physically able to move on land at all.

As well as their anatomical adaptations for swimming and feeding, marine mammals are adapted to coping with the potentially high heat loss that comes from submergence in cold water. Large body size reduces their relative surface area and therefore rate of heat loss, a relationship which explains why the smaller cetaceans, the dolphins, tend to live in warmer waters than those the whales inhabit. A covering of hair is less effective for insulation in the sea than on land because the thickness of the layer of air it traps is reduced as the animal dives and the pressure of the surrounding water increases. Fur seals are covered by a pelt consisting of thick, woolly underfur overlain by long guard hairs, but even this is augmented by a thin layer of fatty blubber beneath the skin. Other seals have replaced all but a thin coating of long hairs by several inches of blubber, while the two permanently marine groups lack hair altogether, apart from sensory bristles around the snout of sirenians. In the largest whales the blubber is as much as 50 cm in thickness. Heat is also saved by how the blood vessels serve the blubber and the flippers. The arteries carrying blood to the outer surfaces are surrounded by the veins that return it. This creates what is called a *counter-current exchange system*, in which a good deal of the heat from the warm outgoing blood passes directly into the cooler returning blood before it gets to the surface and is lost. During periods of activity, or when in warmer water, the animal may need to increase the rate of heat loss, and this can be done by deliberately increasing the flow of warm blood to the vessels close to the surface. The risk of overheating is particularly serious for fur seals when they are moving rapidly overland. There is an old belief that fur seals being chased over the ice by hunters would often die suddenly of a heart attack due to fear. However, what in fact they succumbed to was overheating because they were physically incapable of losing fast enough the extra heat produced by their muscles during the chase.

The nitrogen in the air carried within the lungs of a diving mammal starts to dissolve in the blood as the pressure increases with depth. As the animal returns to the surface, the nitrogen comes out of solution and forms small bubbles of gas. Blockage of fine blood vessels by these bubbles can cause the potentially fatal *decompression sickness*, which means the time of the dive cannot be extended by simply enlarging the lung capacity to take more air down. In fact, cetaceans breathe as much air as possible out of their lungs before diving, and what is left passes into the trachea from where it cannot be absorbed into the blood. The characteristic blowing of a recently surfaced whale is the expulsion of this remaining stale air from the nostril before a breath of fresh air is taken. The length of the dive of a marine mammal is nevertheless greatly increased by its ability to carry large amounts of oxygen already chemically bound to the oxygen-carrying molecules, *haemoglobin* in the blood and *myoglobin* in the muscles. Both these molecules occur in far greater concentration in marine than in terrestrial mammals. There are about twice as many of the haemoglobin containing red blood cells in the blood, and up to nine times the concentration of myoglobin in the muscles. Furthermore, the oxygen that is carried is used to greater effect by reducing the heart rate, a phenomenon called *bradycardia*, and lowering the volume of blood that flows to most of the body apart from the essential organs, these being especially the brain and heart. By these means, and also by being able to withstand a higher buildup of lactic acid in the tissues than can other mammals, an active dive can last up to half an hour for larger seals, an hour for smaller whales, and as long as two hours for the largest whales.

The pinnipeds are the least adapted for aquatic life of the three marine groups because they can still move on land, if not very effectively, using their four legs. In the eared seals such as the sea lion (Figure 27(b)), the front legs are used as paddles for swimming, while the back legs trail passively behind. However, on emerging onto rocks or shores the back legs are turned forwards, and the animal walks using all four legs in an ungainly waddle. The rear part of the body remains on the ground while the fore limbs act as a pair of crutches holding the front part up. Despite the apparent

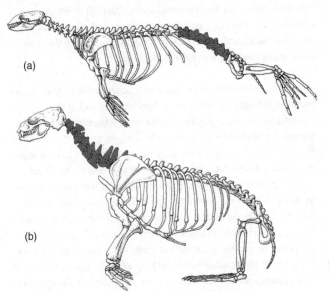

27. **Pinniped mammals: (a) skeleton of true seal, showing hind limbs permanently backwards, and lumbar vertebrae the strongest for hind limb propulsion; (b) skeleton of eared seal, showing hind feet can turn forwards on land, and anterior vertebrae the strongest for fore limb propulsion.**

inefficiency, a sea lion can travel quite long distances at a speed faster even than humans. The earless, or true, seals (Figure 27(a)) are quite different. They cannot bring their back legs forwards at all, but use them and not the front legs for swimming. They hold the soles of the large, webbed feet face to face, and move them together from side to side like a fish's tail, using the flexible lumbar region of the vertebral column. The front legs are used for steering, and especially for manoeuvring when chasing prey. On rocky outcrops or ice floes, where they come out of the sea for safety and to breed and suckle their young, true seals can only progress by a very inefficient sort of flapping of the whole body. The walrus is the third kind of pinniped. On land it uses all four legs for locomotion like the eared seals, although given the massive size of its body, progression is slow and ponderous. On the other hand, walruses swim in a manner more like the true seals, using the hind legs held backwards.

Fish make up the diet of most pinnipeds, although they are never averse to taking invertebrates such as lobsters and, if the opportunity arises, penguins—and even the young of other seals. Some have tended to specialize in feeding on one particular food. For example, the crabeater seal, contrary to its name, feeds mainly on planktonic crustaceans called krill that it collects by filtering the water between its teeth. Each cheek tooth is narrow from side to side and has long, pointed cusps which interdigitate, uppers between lowers. Sea water is passed through the mesh this makes, leaving the krill behind in the mouth. The main food of the predatory leopard seal consists of other seals, penguins, and sea birds which they capture with the help of their large canine teeth. Many pinnipeds will eat molluscs, but the walrus is the specialist consumer of these, especially bivalves such as mussels and clams, which are detected by the long, sensitive tactile hairs on the snout (Figure 28). Despite common belief, walruses do not use their long tusks for collecting molluscs, but rather they dislodge them or dig them up using the toughened horny skin that forms the lugubrious looking 'moustache' along the upper lip (Figure 28). They can also squirt a powerful jet of water from the mouth to excavate deeper lying items. The tusks are mainly for social display, but they can also be used as weapons and to assist movement over ice, using them like ice picks.

Most of us are not very familiar with the Sirenia, perhaps because they are only found in shallow coastal waters and river systems in the tropics. The four species of these strictly herbivorous mammals are the dugong or sea cow of the southern Pacific Ocean and three manatees, the West Indian, the West African, and the freshwater Amazonian of South America. A fifth, Steller's sea cow, was a good deal larger than these, being up to 7.5 metres in length

28. Walrus face showing the sensitive vibrissae and muscular moustache.

and an impressive 10 tonnes in weight. It lived in the north Pacific until its extinction in the mid-18th century at the hands of hunters, who found it an easy source of food. Like whales, sirenians (Figure 29(a)) have no hind limbs, and the forelimbs are short flippers used for manoeuvring. They swim by moving the horizontal tail flukes gently up and down.

Living in tropical waters, having a relatively large, compact body, and a layer of blubber under the skin mean that loss of heat is not such a problem, and the metabolic rate can be low to save on food. Furthermore, living as they do in shallow waters sirenians largely lack predators and spend most of their time drifting in a rather leisurely manner around the great areas of sea grass on which they feed. Their main sense is touch, using the very sensitive *vibrissae*

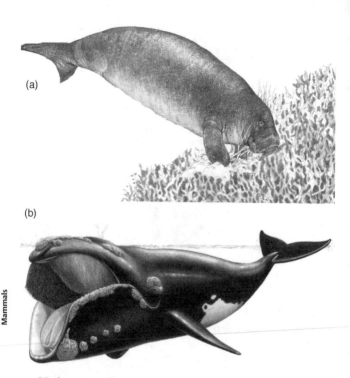

29. Marine mammals: (a) dugong feeding on sea grass; (b) baleen whale with mouth open, ingesting plankton.

around their muzzle and face. Their eyes are small and protected in folds of skin, and their sense of hearing is not particularly well-developed, unlike that of the whales. There is some sound communication between individuals, but apparently little more than simple warning signals to the group and mothers' calls to their offspring.

The upper lip at the front of the short, heavily built head is muscular and mobile for collecting the food, often by digging out the nutritious rhizomes of the sea grass. Manatees do not have any front teeth, but the cheek teeth are large with complex cusp patterns. In surprising contrast for such an otherwise similar animal, the dugong has no more than two or three simple, cylindrical cheek teeth and it uses tough, horny pads for chewing instead. The sirenian intestine is extremely long, and includes a large caecum fermentation chamber containing the microorganisms necessary for digesting the cellulose in their exclusively plant diet.

The reproductive rate of sirenians is very low, another manifestation of their sheltered, unpressurized lifestyle. Female manatees do not reach sexual maturity for several years, after which they only produce a single calf every two years.

There are about eighty-five species of cetaceans, far more than there are sirenians, because whales and dolphins can live in deep as well as shallow waters, and cold as well as tropical regions. A few cetaceans live in freshwater rivers: for example, the Ganges, Indus, and Amazon river dolphins; and, until a few years ago when it was declared extinct, the Yangtze river dolphin of China. All cetaceans are carnivorous. Their diets range from fish, seals, sea birds, and squid in the toothed whales, to the prodigious numbers of krill on which the great baleen whales feed (Figure 29(b)).

Cetaceans are beautifully adapted for swimming. The body is torpedo-shaped, starting at the front with a narrow, elongated skull and shaped down the body with a thick layer of blubber beneath the skin. The nostrils combine to form a single blowhole on the top of the head. There is no outward sign of hind legs, although the remains of tiny pelvic bones associated with the reproductive organs do lie within the body wall. The front legs have evolved into small, smooth paddles for steering, as the animals propel themselves by horizontal tail flukes beating up and down by means of muscles of the vertebral column. Another feature contributing to how well they can swim is the elastic property of their skin, which helps to keep the water flowing smoothly over their body surface, and so reduces the resistance of the water to movement. Speeds of up to 30 knots have been recorded for killer whales, and the efficiency of locomotion of the larger whales allows for migrations of thousands of kilometres. The efficiency also allows them to make very deep dives, for example a sperm whale has been recorded 2,000 metres below the surface of the sea.

Cetaceans possess very large brains relative to their body size, that of the dolphins being second only to humans, and the cerebral cortex is as highly folded as in higher primates (Figure 6). Hearing, particularly echolocation, provides the main source of information about the world that cetaceans have, because neither vision nor smell are much use in the sea beyond a very limited range. Their large brain is associated with a high degree of social interaction, again second only to higher primates. Communication between individuals is vocal, with a wide variety of whistling and clicking sounds for recognizing one another, and for organizing cooperative actions, such as hunting and migrating. Their hunting cooperation is at a very high level, the orcas, killer whales, for example, acting every bit as much as an integrated pack as wolves do. A group of dolphins can corral a shoal of fish and drive it towards the surface by circling around and beneath it in ever tighter formation, before sweeping through the now concentrated ball to grab their prey. Another technique observed in the bottle-nosed dolphin is to swim in a coordinated line making a joint bow wave, and use it to drive fish onto an adjacent muddy shore. With their own momentum, they follow the fish ashore, where they can easily take them before wriggling back into the water. Foraging activities like these can only work with carefully organized interaction between individual dolphins, and an acute awareness of the details of the immediate environment.

Odontocetes are toothed whales, which, as their name indicates, still have a dentition. In most of them it consists of a row of simple, pointed teeth with a single root, suitable for spearing fish which are very easy to swallow and digest, and do not need to be chewed. The number of teeth varies from as many as sixty small ones in each jaw in the spinner dolphin, which feeds on fish at depth, to a mere dozen or so much larger ones in the killer whale, which can take large prey such as seals and young dolphins. The narwhal has the oddest dentition, with a single left tooth in the male in the form of a spiral tusk half as long as the body. It is used as a weapon and also as a secondary sexual character to attract the mostly tuskless females.

Odontocetes detect their prey by echolocation. A stream of air passes over a structure near the blowhole called the *phonic lip* which generates ultrasonic sound waves. These are focused by the *melon*, a fatty body in the forehead that acts as an acoustic lens, and the resulting sound is emitted as a narrow beam in the same way as a torch emits a beam of light. Any sound waves that are reflected back, such as from fish, are received by an oil-filled channel in the lower jaw and detected by the ear. The sensitivity and localizing ability of the system are amazingly high, and objects even several kilometres away can be sensed.

About half the odontocete species belong in the family Delphinidae, the dolphins, which range in size from the West African heaviside dolphin, which is little over 1 metre long, to the 7-metre-long orca. The sperm whale is by far the largest odontocete, males of which have been recorded at 18 metres in length and over 57 tonnes in weight. They are noteworthy for the spectacularly large, barrel-like head filled with a waxy oil, spermaceti, a substance once much sought after by whalers. We are not quite certain, but this oil probably plays a role in echolocation at the considerable depths at which sperm whales feed. The lower jaw is slender and has a row of sharp, recurved teeth, but there are no upper teeth. The main food of the sperm whale is squid, including giant squid that occur at great depths. To collect them, the whale dives vertically down to 1,000 metres or more, where it can remain feeding for over an hour at a time.

The giants not only of whales but of the entire animal kingdom past and present are the Mysticeti or baleen whales. Even the smallest, the pygmy right whale, is about 5 metres in length and weighs 3.5 tonnes; although most are substantially larger. The female blue whale is largest of all; one specimen has been estimated to have weighed 199 tonnes, and another measured 33.58 metres in length. These enormous masses, compared to terrestrial mammals, are possible because of the support of the weight in water, and also because their planktonic food is virtually limitless in abundance. The common name refers to the baleen or 'whalebone', which hangs down as sheets from the upper jaw in place of teeth. It is not bone at all but the hair protein keratin, and each sheet is fringed with fine hairs along the bottom edge creating a huge filter for removing plankton from the sea water. Some mysticetes such as the blue whale have a highly distensible throat shown externally by longitudinal folds. They feed by expanding the mouth cavity to take in a huge volume of sea water and then raising the massive tongue to force it out between the baleen sheets. Krill, which make up the bulk of the plankton, along with any small fish are trapped on the baleen sheets and licked off with the tongue. Another technique, used for instance by the right whale, is simply to swim near the surface of the water with the mouth open so that the sea water is passively forced between the baleen plates, and again the trapped krill are licked off and swallowed.

The adaptive importance of the whale's size lies in the smaller surface area in relation to the volume of the animal, so the whale cools down more slowly in cold waters, further helped by the enormous thicker layer of blubber that a larger whale can cover itself with. Furthermore, the larger the whale the greater the amount of oxygen that can be stored in the myoglobin of the muscles to increase the length of the dives between having to surface and take breath.

In keeping with their large body size, baleen whales range over huge areas of the seas. Annual 3,000–4,000 km migrations occur, between wintering in the lower latitudes where the water is warmer but the food less abundant, and spending the summer in the rich polar feeding grounds. These huge whales are not easy to study, and it is still uncertain whether any mysticetes use echolocation. Certainly there is nothing as elaborate as the odontocete system, but it is possible that large-scale features such as water depth and coastal outlines may be perceived by reflected sound waves. Vocal communication amongst individuals is, however, well-developed and can occur over hundreds of kilometres. Courtship by males, most famously the humpback whale, includes songs that are individually recognizable and responded to by females, while migrating whales presumably use vocalization to maintain the coherence of their group.

Chapter 9
Flying mammals

Being able to fly opens up a whole new range of habits and lifestyles. A flyer can seek out new sources of food such as fruits and seeds from the tops of trees, and airborne insects. It can avoid ground-based predators, and rear its young safely out of their reach. But flying is the most demanding form of locomotion, because it takes a lot more power than walking or swimming. Like an aircraft, a flying animal's wings have to be long and narrow to produce a high enough lift force, but unlike an aircraft the wings have to act as a propeller at the same time. To do this they need to be moved rapidly up and down by powerful muscles that consume a large amount of metabolic energy. We should not be surprised that active animal flight has evolved so rarely, and in only a single mammalian Order, the Chiroptera. Although bats have nowhere near the variety found amongst the 10,000 species of birds, there are still almost 1,200 species of them, amounting to more than a fifth of all mammals, and they are distributed worldwide, sometimes in colossal colonies.

Gliders

There are a few other airborne mammals, which glide passively using a membrane stretched between the limbs rather than actively by flapping wings. They are arboreal and can go directly from one tree to another with little loss of height, or descend rapidly from tree top to ground without injury.

The colugos of the rain forests of south-east Asia are wrongly, if descriptively, called flying lemurs (Figure 30(a) and (b)). There are only two species, the Malayan and the Philippine colugos, which together make up the very small Order, Dermoptera. They are a little over half a metre in length including the tail, and the gliding membrane relative to body size is the largest of all gliding mammals: when extended it stretches from the neck to the ends of the four long, outstretched limbs and the tip of the tail (Figure 30(a)). The eyes are forward facing, giving stereoscopic vision for accurately judging distances. A colugo is capable of well-controlled glides of at least 70 metres, and is well able to seek out new sources of food, which consist of the leaves, buds, flowers, fruits, and even the sap of trees. It rarely ends up on the ground; when it does its movements are very clumsy and it climbs back up the nearest tree as quickly as possible. When at rest during the daytime, a colugo folds the membrane around its body and hangs below a branch by its sharp claws, or else lies in the hollow of a tree. The membrane also forms a protective pouch for the single young, which is born at a very immature stage, an arrangement reminiscent of that of marsupials (Figure 30(b)).

Amongst Australian marsupials, several of the opossums glide between trees in the high canopy. Their gliding membrane is stretched between the fore and hind limbs, but does not include the tail, which is only used for steering (Figure 30(c)). The gliding performance can be most impressive. Sugar gliders, for example, have been seen travelling for over 100 metres from the top of a high eucalyptus tree and still able to land on the trunk of another tree. The third group of gliding mammals are rodents, the flying squirrels, of which there are over forty species. They resemble the marsupial gliders in that the gliding membrane stretches between the limbs but not the tail. As true squirrels,

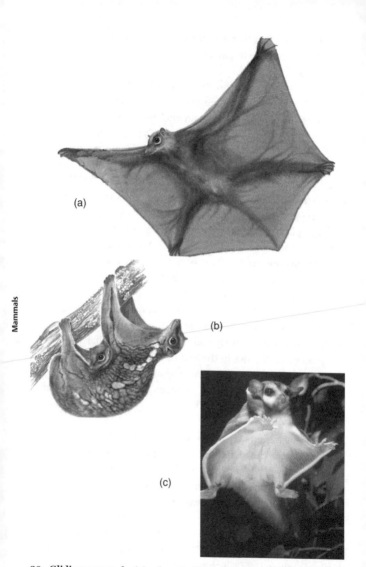

30. Gliding mammals: (a) colugo in flight; (b) colugo at rest with young in membrane; (c) the marsupial sugar glider.

their tail is long and bushy and used for accurately controlling the angle and direction of the glide.

Bats

The gliding mammals evolved so they could move more widely around the tree tops, finding food and avoiding predators such as birds of prey, and we believe that bats evolved from some such gliding ancestor, even though they are not closely related to any of the living gliders. The bat's wing (Figure 31) is made up of a membrane of skin stretched between the shoulder, fore limb and hind limb, and in most of them the tail as well. It is different from a simple gliding membrane because the fore limb is so long, and the second to fifth fingers are enormously elongated to extend and support the outer part and turn it into a proper wing

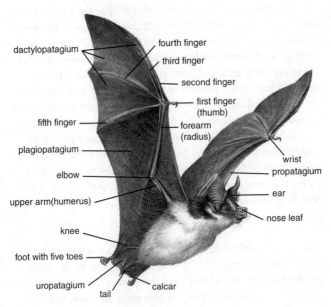

31. Leaf-nosed bat showing wing structure.

shape. It actively beats up and down to create a lift force as effective as that of a bird's wing, and slight adjustments to the fingers control the shape and angle of attack of the wing. In aeronautical terms, the bat's wing tends to have a low aspect ratio (short and broad) which gives the animal manoeuvrable flight, and a high *camber* (curvature of the upper surface) which allows flight to be slower. A typical bat's flight is therefore relatively slow but highly agile, although there is a good deal of variation between different species depending on the habitat. Those living in more open habitats, or undertaking long migrations, have longer, narrower wings for efficient flight at higher speeds than those occupying congested areas such as woodlands, where it is more important to be able to dodge around obstacles.

The power output from the muscles needed for active flight is very high, and it increases disproportionately as the body weight increases. There is a limit to the weight that the wings could support, which explains why bats are small: even the largest species only reach 1.6 kg in weight while the smallest, the bumblebee bat, has a head and body length of 3 cm and weighs less than 2 grams, making it the world's smallest mammal. The very high power of active flight imposes another condition, a very high metabolic rate fuelled by large amounts of food. At times when not enough food is available, to prevent starvation bats can become temporarily torpid, when the heart rate, breathing, metabolic rate, and body temperature all fall dramatically. Torpor can be for regular periods during the daytime when feeding normally ceases, or for longer periods of several weeks of hibernation during the winter season in temperate parts of the world.

Bats are extremely sociable creatures, often to be found roosting in colonies which may number a million individuals, all hanging upside down by the claws of their first fingers. They find safety in trees, caves, or natural crevices within rocky outcrops, and some have taken advantage of human structures, to be found anywhere from roof spaces to underground mines. The advantage an individual bat gains from the colonial habit is in keeping warmer as it roosts close to others during the night. It is also less likely to be picked out by a predator when it is in the midst of so many other individuals.

Chiroptera fall into two groups, the Megachiroptera, or fruit bats, and the Microchiroptera, which are the mainly insectivorous bats hunting their prey by echolocation. Echolocation is a way of creating what we can imagine as a sort of auditory picture of the surroundings, so that even in complete darkness the bat can manoeuvre amongst objects and obstructions such as trees, and detect and capture flying insects. Short pulses of ultrasonic sound are emitted from the mouth, or from the nose in the case of horseshoe bats. Typical sound wave frequencies are in the range of 20–60 kHz, which are equivalent to wavelengths short enough to rebound as an echo off an object as small as an insect, but long enough to be effective over a distance of several metres. The sound is emitted in brief pulses up to around 30 milliseconds long, and the returning echoes are then detected by the extraordinarily large, sensitive ears.

The mode of feeding in bats that we are most familiar with involves flitting erratically around in the night air capturing airborne insects, a behaviour seen in nearly all the 300 or so vespertilionids, the largest group of microbats. A number of other bats are carnivores, for example the 100-gram greater false vampire bat of Australia, which is the largest microchiropteran. It feeds on other bats, reptiles, and amphibians which it can detect by normal hearing as well as by echolocation. Hunting is from its perch on a tree, to which it returns with its captured prey to ingest it. The true vampire bats are South and Central American, where they feed on the blood of birds and mammals. Unusually for bats, their prey is detected by smell not sound, and it is often approached along the ground. Heat sensitive receptors on the nose detect one of the prey's superficial blood vessels, and extremely sharp, pointed incisor teeth scrape away a small patch of skin to expose it. The blood is lapped up by the tongue and there are anticoagulants in the saliva to prevent clotting. Vampire bats exhibit a behaviour called blood-sharing, which is one of the most remarkable examples of altruism or mutual cooperation amongst adult mammals. An individual within the roosting colony will regurgitate blood from its stomach and donate it to a hungry bat, most often, though not always, one of its relatives.

Fishing is another specialized feeding adaptation. The fisherman's bats of Central and South America use their echolocation to detect ripples on the water created by a fish near the surface. They fly down and catch the fish with their very long legs, and carry it away in a cheek pouch.

A number of the New World family of leaf-nosed bats have completely abandoned their ancestral ways and become herbivores, eating fruit or in some cases the pollen and nectar of flowers. They are important as pollinators of certain plants, including the maguey plant from which the Mexican alcoholic drink tequila is made.

The second group of bats are the Megachiroptera, the fruit bats, which are often called flying foxes because of their long, rather dog-like faces and normal sized ears. They are tropical animals found in Africa, Southeast Asia, and Australia, and on many Pacific islands. Unlike microchiropterans, fruit bats are unable to echolocate and they seek their food by vision, helped by a degree of colour sensitivity. The largest bats of all are members of this group: the Indian flying fox weighs 1.6 kg and has a wingspan of 1.7 metres. Their main food is fruit, although flowers too are readily consumed. Fruit bats are also important pollinators, and agents of seed dispersal, and several species of tropical plants have evolved large, conspicuous flowers and excessive nectar to attract them. Some species of flying fox occur in prodigious numbers, and few sights in nature are more awesome than over a million individuals roosting together and completely covering all the branches of an area of the forest. Not surprisingly, they can be a major pest in some tropical fruit growing regions, being capable of destroying an entire crop.

Chapter 10
Primates

The Order Primates is the only mammalian Order to have lent its name to a widely recognized discipline of biology, complete with university courses, professorships, and dedicated textbooks: *primatology* is the study of the biology and evolution of humans and their primate relatives. It has revealed to us the extraordinary story of how a group of small, rather insignificant tree-dwelling mammals living sixty million years ago eventually evolved the highest level of expression of the mammalian characteristic of adaptable behaviour by means of a large brain. Many people have speculated on why this should have been so. One factor is that living in trees imposed a selection pressure for three-dimensional awareness of a complex environment, using binocular vision to judge distances with great accuracy. There was also a need for long fore limbs and hands capable of grasping, to hold onto branches and pluck fruit, leaves, and insects from them without falling. A third factor was probably the evolution of greater levels of social cooperation amongst individuals for warning of predators of what were relatively defenceless, very visible, daytime feeders.

Traditionally the living primates were classified as a series of increasingly more evolved taxa believed to illustrate the broad picture of human evolution. The least evolved, or most primitive, group were the prosimians, which are the lemurs, lorises, bush babies, and tarsiers. They were followed successively by the monkeys, the gibbons or lesser apes, the great apes, and finally the humans. Nowadays, however, we have learned that these groups do not reflect the true evolutionary relationships at all well. To start with, the tarsiers are more closely related to the higher primates than to the other prosimians; then the old world monkeys of Africa and Asia are more closely related to the apes than they are to the new world monkeys of South America; amongst the apes, the orangutan is less closely related to humans than are the gorilla and chimpanzees; even between these two African apes, the chimpanzees share a more recent common ancestor with humans than does the gorilla.

Lemurs, lorises, and bush babies

The least evolved of the living primates are the lemurs of Madagascar, the lorises of tropical Africa, India, and Southeast Asia, and the bush babies of sub-Saharan Africa. They all belong in a group called Strepsirhini, whose most distinctive characteristic is a tooth comb made of the narrow, elongated lower incisor teeth, which is used in grooming and feeding. We cannot mistake them for anything but primates because of the forward-facing eyes overlooking a shortened snout, the brain which is quite large for a mammal, the flattened finger and toe nails, and the square crowns of the molar teeth. But they still have several characteristics of the primate ancestor. The snout is only modestly reduced and the brain modestly enlarged, and there is a *rhinarium*, a moist pad at the end of the nose that is seen in many other mammals such as dogs. Strepsirhine social behaviour is generally simpler than in most other primates. In fact many are virtually solitary for much of the time, some live only as pairs, and while others do live in groups, these are small and consist of a similar number of socially equal males and females, quite unlike the complex, alpha-male dominated hierarchies more typical of primates.

We do not know exactly why the lemurs are entirely restricted to the island of Madagascar; perhaps there was an accidental dispersal of their ancestors from mainland Africa at an earlier stage in primate evolution. What is clear though is that there have never been any other competing primates there, because several of the thirty-five living species of lemurs evolved to fulfil roles that monkeys and apes perform elsewhere. The ring-tailed lemur, for example, leads a monkey-like existence as a forest dwelling, arboreal herbivore, although it is also perfectly adept at moving on the ground using all four of its equal-sized legs. It feeds during the day on fruit, flowers, and leaves, and is not averse to catching insects and even small vertebrates.

The indri and the sifakas (Figure 32(b)) follow a gibbon-like lifestyle. They stay high up in the forest canopy as much as possible, where they travel using the powerful hind legs to leap between branches as far as 10 metres apart. Also like gibbons, they communicate with one another by singing calls. These are the largest living lemurs, weighing up to about 9 kg. However, until their extinction soon after the arrival of humans in Madagascar a few thousand years ago, there were ground dwelling giant lemurs that weighed up to 160 kg, the size of the largest male gorilla. At the other extreme, the pygmy mouse lemur is the smallest primate of all, weighing less than 40 grams. Mouse lemurs (Figure 32(c)) are only active at night, feeding on a high energy diet of insects, fruit, and tree sap, and readily entering a state of torpor when food is scarce. They can be compared in their mode of life to the bush babies of mainland Africa. The aye-aye (Figure 32(a)) is the strangest of the lemurs, looking more like a bushy-tailed rodent than a small primate. Its extraordinary feeding behaviour can be compared to that of a woodpecker. The incisor teeth grow continuously like those of rodents, and are used to gnaw into dead wood, from which the aye-aye extracts insect larvae using its very elongated fingers.

The lorises (Figure 32(d)), bush babies or galagos, and pottos are all small, fully arboreal strepsirhines that move around using all four limbs to maintain a grip on the branches. Galagos are excellent

32. Primates: (a) aye-aye; (b) Verreaux's sifaka; (c) grey mouse lemur;
(d) slender loris; (e) tarsier; (f) mandrill; (g) spider monkey;
(h) gibbon; (i) orangutan; (j) gorilla; (k) chimpanzee.

leapers using their larger hind limbs, and have a long bushy tail for balance. In contrast to them, the lorises are very slow moving and are almost tailless. Their limbs are all the same length, and they use them to travel from branch to branch, grasping them with a grip so tight it is almost impossible to break. The main food of all the members of this group of strepsirhine primates is insects, including poisonous kinds of caterpillars in the case of the lorises. The larger species readily consume fruit as well.

Tarsiers

The tarsiers (Figure 32(e)) are only found on the islands of Southeast Asia. They were once classified with the lemurs, galagos, and lorises even though they have several characteristics in common with all the rest of the primates. For example, the end of the nose is hairy rather than moist, and they have a multi-layered placenta. Tarsiers are nowadays included with monkeys and apes in the Haplorhini, a relationship established beyond question by molecular evidence. Despite this, tarsiers do rather resemble galagos, both having long—in the tarsier's case extremely long—hind legs, and a long, hairy tail. Although no more than about 12 cm in body length, a tarsier can leap a prodigious distance of several metres, even a mother carrying its young clinging onto the underside of her belly. They rely on long, dexterous fingers and toes with swollen tips to catch onto the trunks and branches of trees as they land. Tarsiers' eyes are the largest of any mammal, relative to their body size, and they have excellent night vision, helping them in their search for insects and small vertebrates, which are then captured by the delicate fingers.

New World and Old World monkeys

All the other primates, the monkeys, apes, and humans, make up the Anthropoidea. They have a larger brain than tarsiers, with a monkey's brain being three times the size of that of an average mammal of the same body size, mostly due to an increase in the neocortex. The enlarged brain is responsible for the enhanced learning and cognitive ability, and the more complex social interactions that are characteristic of anthropoids. One expression of this is that they can individually recognize a greater number of members in their group, and act appropriately towards each one, such as appreciating their social status in a dominance hierarchy. Another is the number of vocalizations used that have particular meanings. For example, vervet monkeys have over twenty distinct calls, including different warning calls for different predators such as eagles, leopards, or snakes, and for indicating different kinds of food. They have a range of social calls between individuals relating to their rank, mood, and intention, all designed to elicit an appropriate response from the recipient. There are also at least sixty distinct physical gestures they make to one another.

The main sense of the anthropoids is their vision, including red/green/blue colour vision that is especially useful for finding ripened fruit on the trees. Their colour vision has also led to the use of coloration for sexual signalling in some species, for example the bright blue and red facial pattern of the dominant male mandrill, and the bright pink buttocks that a female baboon displays when she is receptive to a male. Indeed monkeys include the most colourful of all mammals. Compared to the strepsirhines, the sense of smell of anthropoids is quite poor and of much less importance than vision.

The evolution of the anthropoids took place in Africa, but over thirty-five million years ago one species managed to reach the then isolated island continent of South America. Perhaps a few individuals were carried on a mass of vegetation torn off the African coast in a storm, and by chance floating all the way across the then much narrower Atlantic Ocean.

Since then this new arrival evolved into the New World branch of monkeys, known as the Platyrrhini on account of a characteristically broad nose with widely spaced nostrils. Today there are about eighty-five species in South and Central America, ranging from tiny marmosets and tamarins weighing as little as 120 grams to the larger, more diverse cebid monkeys such as the 10 kg woolly spider monkeys. Most New World monkeys have a long, prehensile tail capable of grasping branches and acting like a fifth limb. Spider monkeys (Figure 32(g)), for example, move with great agility and speed through the forest canopy in search of the ripened fruit and tender leaves that make up their diet. The howler monkeys are another remarkable member of this group. Their voice, amplified by a bony resonating box in the throat, is the loudest noise made by any terrestrial animal, for it can carry up to 3 miles through the forest. The main purpose of the roaring is to indicate the troop's territory, which has to be large enough to contain sufficient of the tender, nutritious leaves that make up most of the diet. Sound is a much more efficient way of defending territory than the more usual method of patrolling borders and fighting off rivals. The marmosets have specialized in eating the gum which is exuded by many trees as protection against insect attacks. To collect it, they use their sharp claws to cling to the tree trunk, and gouge out a shallow hole with their sharp, chisel-edged incisor teeth. Gum is an important year-round resource for these primates that are too tiny to forage very widely.

The forty-five or so species of Old World monkeys of Africa and southern Asia make up the Cercopithecoidea. Most of them are tree-dwelling fruit and leaf eaters, living in tropical forests where food is abundant all the year round. The colobine monkeys, such as the langurs and leaf monkeys of Asia and the colobus monkeys of tropical Africa, are the most efficient feeders. They have an enlarged stomach divided into several regions full of bacteria for digesting the cellulose of the plant cell walls. The cercopithecine monkeys include a number of species that spend more time on the ground than in the trees, and these tend to have the most versatile feeding strategies. Vervet monkeys occur throughout most of sub-Saharan Africa, where they will consume almost anything, from leaves, flowers, fruit, and seeds, to insects, eggs, and nestlings of birds—and even small mammals. Troops of them have become adept at raiding human crops and even pestering tourists for food. Although making use of trees for safety most of the time, vervets happily cross open spaces. The baboons are even more habitually ground-dwellers, walking and running on all four equal-length legs. They can even run at a fast enough speed to catch mammalian prey such as hares and juvenile antelopes.

Apes

There are not very many apes, but they are of special interest to us because they are our closest living relatives, and it is hard to avoid thinking about the similarities. There are two families. The Hylobatidae, or lesser apes, are the dozen or so species of gibbons, which occur from northeastern India and southern China through to the islands of the Malay archipelago. The Hominidae, or great apes, are the orangutan of Sumatra and Borneo plus the chimpanzees and gorilla of Africa—and also *Homo sapiens*.

Several of the features of the apes' body evolved in the first place for a special mode of arboreal locomotion called *brachiation*, which is swinging beneath the branches rather than walking on top of them as do monkeys. The arms are longer than the hind legs, the chest is broad, the shoulder girdles are mobile, the hands and feet are grasping hooks, and there is no tail. The gibbons have taken this ability to extremes (Figure 32(h)). With the help of very long arms and almost equally long legs, they undertake acrobatic travel over long distances by rapidly swinging from branch to branch, and leaping between gaps in the forest canopy. Being quite large primates, they need to be able to forage widely like this to find enough of their diet of ripe fruit to satisfy their needs.

A family group of gibbons, usually consisting of a monogamous pair together with two or three of their offspring, advertises its presence by amazingly elaborate singing which can be heard several miles away. A song consists of series of ascending notes, trills, and crescendos, and each individual can be recognized by its unique version. Typically, the male sings his song daily at dawn and the female her very different song mid-morning. Singing like this serves several functions, including identification amongst members of a group and between those of nearby groups. The male song also discourages potential rival males from approaching his female. The female song, which is sometimes accompanied by the male as a duet, is used to advertise the group's territory and discourage invaders from other groups.

Even the largest gibbon, the siamang, only reaches about 10 kg in weight. A male orangutan (Figure 32(i)), the 'old man of the woods', can be as much as 90 kg, and at this weight the carefree brachiating of the gibbons is impossible. While still almost entirely arboreal, the orangutan's movements are slow and cautious, and it always holds on by at least two of its hook-shaped hands or feet as it moves from branch to branch. To cross a gap between trees, rather than leaping, it swings on its own branch until it approaches closely enough to an adjacent one to be able to grasp it. In this laborious manner, an orangutan will go scarcely more than 1 km a day, which may explain why they do not defend territories as do the gibbons. Their food consists mainly of fruit, though young leaves, seeds, bark, and even occasionally insects are eaten.

The African apes are also large primates, male chimpanzees up to about 40 kg and male gorillas 180 kg. Both are capable tree climbers and, in the case of the chimpanzees, most of their life is spent there, feeding by day and resting by night. Gorillas are far less frequently to be found in trees, which they climb only to collect fruit when it is available. For moving on the ground, both have a mode of locomotion called *knuckle-walking*, in which they use the soles of the hind feet in the normal way, but the fingers are bent back and the hands carry the weight on the knuckles rather than on the palms. This allows the animal to walk and run on all fours without their hands having had to lose the important capacity to grasp.

Gorillas (Figure 32(k)) are found in only two regions of equatorial Africa, the western gorilla in the swamps and forests around Cameroon and the eastern gorilla up to an altitude of 3,790 metres in the vicinity of Uganda. Contrary to their unwarranted reputation for being fierce, gorillas are normally unaggressive and are exclusively herbivorous, preferring the higher energy source of fruit, but capable of surviving when necessary by spending long periods of the day browsing on leaves and stems. To this end, the gorilla's molar teeth are large, and the jaw muscles powerful. The gorilla social structure consists of stable harems of five to twenty females plus a single, much larger and dominant male, the 'silver-back', who uses his sheer strength and very large canine teeth to protect the group from predators such as lions, and to chase off potential rival males.

There are two species of chimpanzee, the common (Figure 32(k)) and the bonobo or dwarf chimpanzee. Even though anatomically chimpanzees are quite similar to gorillas and orangutans, molecular evidence shows that, in fact, we are their closest living relatives. The chimpanzees' habitat is the equatorial rain forest and savannah of central and western Africa, where they live in groups of anything from fifteen to 150 individuals. Using their very long arms for swinging, and elongated fingers for holding on, they are agile tree climbers, where they spend most of their time feeding during the day and sleeping at night. However, unlike the gibbons and orangutan, when going any distance they mostly travel over the ground. Their food is more mixed than that of the other apes, and whilst fruit and other parts of many different plants form the bulk of their diet, they also consume bird nestlings, and they can capture mammals as large as monkeys, bush pigs, and small antelope. They achieve this by well-organized communal hunting. Typically, some of the members of the group surreptitiously go ahead of the potential prey to block off its escape routes, and once in place the others chase it through the trees or along the ground, with much shouting, screaming, and stick throwing, until it is driven onto the ambushers, who dispatch it with their strong canine teeth.

A great deal has been discovered about chimpanzee behaviour in the wild, starting with the pioneering observations of Jane Goodall in the 1960s. Furthermore, their smaller body size compared to that of gorillas and orangutans makes them far easier to maintain in captivity, where their learning ability and potential language skills can be studied experimentally. As more is discovered about the mental capacity of chimpanzees, the more human-like their behaviour seems. The average brain size is about 400 ml, which is almost a third that of a human brain, and is the largest relative to body size of all primates (Figure 6). It is particularly characterized by the size and extent of the folding of the neocortex, similar to though not as extreme as that of humans.

Chimpanzee social groups are quite fluid, as individuals join and leave from time to time, but nevertheless they are highly organized. There is a linear male hierarchy of dominance topped by a single alpha male, and all males are dominant to all females, which in turn have their own hierarchy. Communication amongst individuals is extensive and pretty well continuous. About thirty different calls indicate such things as greetings, alarm, mating inclination, food sources, and what can only be described as states of mind like contentedness, bad temper, aggressiveness, and desperation. An equally impressive range of visual gestures and facial expressions are used to convey friendship, fear, begging for food, and aggression of one towards another, using grimacing, stamping, hand clapping, throwing sticks, or charging. Submission towards a dominant individual is expressed by facial pouting, crouching, or showing the rump. Bonding and communication is particularly strong between the mother and infant, and, as in humans, there is a relatively long childhood. Nursing lasts for about five years, during which the young indulge in child-like curiosity and play; it is a period of intense learning of social and life skills. Also like humans, the maternal bond often lasts a lifetime.

Although good, if noisy, relationships generally prevail within the group, various forms of severe aggression can occur. For example, a male will sometimes kill an infant in order to impregnate a female with his own offspring. At a community level, what looks very like warfare occasionally takes place between two groups, during which ferocious battles ensue, involving the use of sticks and throwing of stones as well as direct physical combat. The conflict may well continue until one of the groups is completely eliminated.

Tool use is widespread and versatile. Sticks may be broken to a suitable size and stripped of bark for use in digging up termites and reaching honey in bee hives. Rocks are often used as hammers and anvils to smash hard fruits. Individuals have been watched chewing up leaves to make a sponge for soaking up water from a puddle. The learning capacity of chimpanzees in captivity has been shown to be phenomenal by the standards of other mammals. Most impressive, perhaps, is that they can be taught to understand the equivalent of as many as 300 different words in sign language, although they are unable to learn to speak at all. There is little doubt from various lines of experimental evidence that chimpanzees are capable of relatively abstract, conceptual reasoning, and also that they are self-aware, and can experience emotions such as happiness and sadness in a manner comparable to humans.

Humans

The idea that we humans are members of the animal kingdom, no different in principle to others albeit possessing our own specific adaptations, took many centuries to take root, even amongst scientists. But as the theory of evolution developed, and pre-human fossils started to come to light, our primate nature grew clearer.

Indeed, the distinction between great apes and humans becomes ever less hard and fast, as our appreciation of the level of intelligence of the latter grows.

The two most important new adaptations to evolve in humans are *bipedalism* and a huge brain of around 1,400 ml capacity. Bipedalism is walking on the hind legs with the body held vertically, and it required several changes to the skeleton. The pelvis needed to be short, the hip joint modified so that the leg was directed downwards, the knee joint straightened, and the feet narrowed. The original adaptive importance of bipedalism has been debated for a great many years. The most widely held theory is that raising the head high gave a better viewpoint above the savannah grasses when moving on the ground, and at the same time the arms were freed for carrying things such as food, weapons, and infants. However, evidence has been growing in recent years for an alternative explanation called the *waterside ape theory*. This is that, at some stage in their evolution, pre- or early humans were partly aquatic, wading in lakes and estuaries, and along the seashore to collect shellfish and water lily bulbs with their hands. This would explain why the body is held so vertically, and why the thumb opposes the fingers. It also accounts for some of the less easily explained characteristics of humans, such as the replacement of hair by a fatty insulation layer, the existence of a diving reflex, and the swimming ability of very young infants—all features matched by marine mammals but not by any other primates. We also have direct evidence from some early human fossil sites that shellfish and fish did play a major role in the diet of at least some early pre-human groups. Whatever the original purpose of bipedalism was, it certainly was not for fast running, because we are not capable of the speeds typical of most mammals of our size. We do, however, have the endurance to keep on running for long distances.

The evolution of a huge brain required several changes to the anatomy of the skull, giving it its distinctive human form. The cranial vault is greatly increased to house the brain. To compensate, the muzzle is extremely short, and the teeth and the jaw muscles are reduced in size, suggesting that meat, fish, and fruit made up a larger part of our early diet than did vegetation. The *condyles*, the points of attachment of the skull to the vertebral column, have moved to a position underneath the skull, making it easier to support the weight of the skull on the top of the vertical neck. The human brain is about four times the size of the ape's brain, reflecting our more complex social life as well as our far greater learning abilities and adaptability of behaviour. It is also associated with the uniquely human attribute of language. The apes can recognize and respond to a large number of verbal signals, but these are equivalent to single words rather than sentences: they lack the use of syntax. Human speech can convey vastly more detailed information by joining nouns, verbs, and qualifying words into sentences. These are spoken and understood in the context of a rich background of abstract concepts, acquired knowledge, and memories shared by those in communication.

Although we humans gained our characteristics by evolution, it is not always helpful to regard ourselves simply as another species of primate with certain anatomical and behavioural adaptations. We shall consider the consequences the differences have had for the rest of the mammalian world in the following chapter.

Chapter 11

Humans and mammals: the past and the future

The past

We should not forget that the extinction of existing species is just as much a part of evolution as the origin of new ones, because without the first leaving the scene there would be no room for the second. The fate of all species is to disappear eventually and to be replaced by either their own descendants or those of other groups. Mammals are no exception, as we saw in Chapter 4, and over the course of their evolution great changes in the numbers and kinds of species have taken place. Fifty million years ago during the Eocene epoch, global temperatures were much higher than they are today, and that was the time when the greatest number of mammals ever lived. Yet fifteen million years later the world had grown cooler and drier, and over half of them had disappeared. Another fifteen million years later, the climate had improved and the Miocene radiation started, with many new kinds of mammals evolving, only to be followed by the Pliocene decline as temperatures started to fall once again, a phase beginning seven million years ago and continuing on and off to the present day.

The reasons for the climate changes and the extinctions and radiations that coincided with them are complex, but the main underlying cause was a fluctuation in the level of carbon dioxide in the atmosphere. This caused variation in the *greenhouse effect*, which is the increase in the sun's heat trapped in the Earth's atmosphere the more carbon dioxide there is present. Increasing carbon dioxide has another, more direct effect on life because the more there is available to the plants, the more they can photosynthesize and grow, providing more food for herbivorous mammals, which in turn feed their predators.

Seen against this background of natural species turnover, however, the change in the mammalian fauna of the last few tens of thousands of years is different. There has been another large extinction phase of mammals, but unlike earlier phases it has been heavily biased against species of larger body size. Such mammals as mammoths, sabre tooth cats, the giant Irish elk, a giant ape, and a giant hyaena have disappeared. The huge ground sloths, glyptodonts, and notoungulates of South and Central America, and the 1-tonne *Diprotodon* and giant kangaroo of Australia have been lost. Furthermore, this extinction phase has been much more rapid than those of the past, taking only a few thousand years instead of the tens to hundreds of thousands of years of earlier mass extinctions. But the most significant feature of all is that the extinction occurred at different times in different parts of the world. These times coincide, at least approximately, with movements of the human population as it underwent a great expansion and worldwide dispersal as the last ice age drew to a close. In Australia, the extinction period started around 60,000 years ago, in Eurasia around 40,000 years ago, and in America 10,000–12,000 years ago. As recently as 1,000–2,000 years ago, humans arrived on the island of Madagascar, and this date coincides with the extinction of the giant lemurs, and also with the date of the earliest cave paintings there depicting the hunting of these animals.

We can scarcely avoid concluding that human activity was the main cause of this *megafaunal extinction*, even if natural climate change has possibly played some part as well. Archaeological evidence from rock paintings and human artefacts speaks of advances in weapons such as bows and arrows, and in techniques of hunting large animals, for example by using pit traps dug for the purpose. Even though the human population was still extremely low by modern standards, computer models of likely rates of killing by hunting compared to the prey's rates of replacement by breeding show that humans could indeed have driven to extinction the large mammal species that they most hunted, over the course of as little as 1,000 to 2,000 years.

From the very start of the spread of humans, the world's mammals have been irreversibly and mostly detrimentally affected, due to direct exploitation for food and for skins to make clothes and shelter. This hunter-gatherer lifestyle of humans was continued well into modern historic times by many communities—even today it can be seen to a limited extent in a few parts of the world. In the particular case of whaling, for example, this practice was until quite recently the preserve of indigenous, coastal people hunting with primitive weapons from small, precarious craft. By the 18th century, whaling had grown to a commercial scale to satisfy the demand for sperm oil for lamps, candles, and soap; whalebone for corsetry and much else in the pre-plastics age; and whale meat for human and animal sustenance. The invention of sonar for finding whales and of explosive harpoon guns to kill them without risk to the whalers vastly increased the efficiency of the hunt, and several whale species are now so severely reduced in numbers that they face imminent extinction.

Other mammal products for which humans developed a more irrational, but nonetheless insatiable desire were elephant ivory for decorative purposes, and rhino horn and pangolin scales to serve the totally false belief that they possess special medicinal properties. No longer were the skins of mammals valued only for clothing to aid human survival, but species such as beavers in North America, sables in Russia, and koalas in Australia were killed in their millions during the early years of the 20th century simply in the interest of fashion.

The rise of agriculture in permanent settlements led to a rapid increase in the human population, many now living in towns and cities, and this had another major impact on the world's mammals. A small number of species were domesticated, species that happened to combine a nature that is amenable to living in human company with usefulness to society. We know from archaeological and molecular evidence that dogs were the first domestic animals. They originated from wolves between 20,000 and 30,000 years ago in different regions of Eurasia. Wolves and humans use a comparable hunting technique based on group cooperation, and they seek out similar prey. Puppies and young wolves were found to be easily trained to work with a human rather than a wolf pack in chasing and bringing down prey; selective breeding of the least aggressive individuals rapidly led to fully domesticated dogs.

We find evidence of domestication of mammals for another purpose a few thousand years later. Wild species of sheep and goats live in social groups, have a convenient body size, tend not to be aggressive, and easily become habituated to human presence. Herding them is a very efficient and sustainable way of utilizing their meat, milk, and skins, both in nomadic and in settled agricultural communities. The habit arose in central and western Asia around 11,000 years ago. Shortly afterwards, cattle appeared. They were domesticated from the now extinct wild ox or auroch, and have since become the most important meat and milk source worldwide, with more than 800 different breeds alive today. Pigs were the third main group of mammals domesticated for food. They are descendants of the wild boar, and were domesticated independently several times in Eurasia, starting around 9,000 years ago. Their very broad diet would have led to wild populations scavenging for food waste around human habitations, and their palatability as food soon encouraged humans to pen them in and feed them directly, providing themselves with a rich and ready food source.

After dogs for hunting, and goats, sheep, cattle, and pigs for food and skins, the third category of domesticated mammals was for transport. The modern horse was domesticated from the tarpan, or wild horse, around 5,000 years ago by the tribes of the Eurasian steppes. Before long, horses had revolutionized the movement of people and goods, and they had a considerable effect on the conduct of warfare as well. Beasts of burden domesticated in other parts of the world were asses, camels, llamas, and elephants.

This mere handful of domesticated mammal species has had a huge impact on the rest of the world's mammalian fauna, and indeed on its whole biota. Extensive areas of the Earth's most productive land have been given over exclusively for animal grazing, not only existing natural grasslands but, even more damagingly, tracts of pasture created by cutting down forests. Native mammal species, predators, and those potentially in competition with domesticated animals for food are actively excluded. In sub-Saharan Africa for example, cattle play an important economic and social role, and they are rigorously protected from competition for food by elephants, antelopes, and zebra, and from predation by lions and leopards. In the whole of Europe, deer are almost completely excluded and wolves virtually exterminated from grazing lands, as well as from other great areas used for producing crops for human and animal feed.

The future

By the time of the agricultural revolution 8,000 years ago, the human population is estimated to have been five million. It reached one billion by around 1800, and today it is 7.5 billion. Most estimates suggest it will approach ten billion by the year 2050. The problem of habitat loss to mammals due to agriculture, forest destruction for fuel and building materials, and the remorseless spread of urbanization have been added to over the last couple of centuries by atmospheric and oceanic pollution from mining and manufacturing processes, and by global warming caused by the release of CO_2.

About 240 mammal species are known to have become extinct over the last 12,000 years, which is perhaps a surprisingly small percentage of the 5,500 or so mammal species alive today. Over the last 400 years, there has been complete extinction of only a few tens of mammals, including species as diverse as the auroch (1627), Steller's sea cow (1768), the thylacine (1930s), the Australian blue-grey mouse (1956), the Yangtze River dolphin (2006), and the Christmas Island pipistrelle bat (2009). The crisis faced by the world's mammal fauna at this moment is measured less by the extinctions that have occurred than by the severe fall in population numbers of most species. Below a certain critical number of individuals, a species is unlikely to survive for long, because populations fluctuate naturally, and if one is too small it is almost certain that by chance it will fall to zero sooner or later. If the population is suffering from loss of habitat or over-exploitation by humans, its extinction is even closer to becoming inevitable. The current estimate is that about a quarter of all living mammal species are threatened in this way, many severely and probably irreversibly so, and if trends continue as they are, this proportion will rapidly increase.

The strictly rational solution to the conservation crisis is as biologically obvious as it is culturally impossible: a huge reduction, preferably to zero, of the human population. Any realistic approach, however, will have to be a compromise heavily weighted in favour of the demands of humans for natural resources. Nevertheless, if the desirability of conservation comes to be widely accepted, and the necessary degree of self-restraint in human behaviour imposed, a good deal could be possible. By far the single most important practical measure is the maintenance, extension, and rigorous policing of national nature reserves. Enlightened governments of several countries led the way in the late 19th century, such as the establishment of the Yellowstone National Park, the Adirondack Forest Preserve, and other National Parks in the USA, and James Stevenson-Hamilton's creation of South Africa's Kruger National Park. Generally speaking, the larger the reserve the more effective it will be in supporting a diverse habitat and its biota. In this light, an exciting initiative is currently under way. A transnational park called the Great Limpopo Transfrontier Park has been created to link the adjacent Limpopo National Park in Mozambique, the Kruger National Park in South Africa, and the Gonarezhou National Park in Zimbabwe. The Kavango-Zambezi Transfrontier Conservation Area is a similar development, aiming to connect wildlife parks in northern Botswana, southwest Zambia, and northeast Namibia. These have the potential to house, conserve, and facilitate the migration patterns of a rich and diverse indigenous biota. The second most important action after establishing such areas is to protect them effectively against poaching and economic pressures for mining, timber extraction, and agricultural encroachment.

Ultimately, successful conservation has to depend on government and local community commitment, and on educating and enthusing the next generation, our children, on how desirable and important conservation is. An immediate, practical approach to this end would be to make sure that the substantial income available from modern ecotourism goes to local community organizers rather than to foreign operators. For example, in 2015 Namibia employed 25,000 people and earned about US$1.7 billion from tourism, which represented 15 per cent of the country's total GDP. Unfortunately, in certain other parts of the world containing rich natural faunas, neither the start-up funding nor the political will are yet available for developing and protecting natural areas to this level of local economic benefit. Conservation efforts are heavily trumped by socio-political pressures for human land use, a process frequently exacerbated by governmental inefficiency and corruption.

An alternative to the blanket conservation of large protected areas is to focus resources on particular species, usually the more iconic ones for which it is easier to raise funding. The giant panda, for example, while still listed as endangered, is recovering from a catastrophic population fall to 1,200 individuals and is now approaching 2,000 thanks to intensive protection from poaching and the conservation of patches of suitable habitat by the Chinese authorities. Another successful example relates to the Arabian oryx, which was virtually exterminated from the wild by over-hunting. However, an intensive programme of breeding captive animals has produced enough individuals to release them carefully into a reserve in Jordan, where their numbers have since risen to over 2,000. While captive breeding can work well with individual species, it is of course extremely limited in how many threatened mammal species can hope to be saved in this way.

We have to accept that preventing the extinction of all mammal species is an impossibly optimistic aim, and so we must be objective in working out our priorities in this regard. The first step is to expand the worldwide research effort presently underway to gain as much knowledge as possible about the status of all mammal species and their habitats. We can then agree on suitable criteria for deciding where our limited resources can most effectively be allocated. The criteria should include such factors as how much at risk of imminent extinction and how evolutionarily distinctive a given species is. A simple example in use is the EDGE score, which is decided on a species' evolutionary distinctiveness (ED) and how globally endangered (GE) it is according to the International Union for the Conservation of Nature (IUCN)'s Red List. The risk of extinction of a species does not, however, depend solely on its present population size and rate of decline, but also on factors like whether it is an island species, a species with a fragmented habitat, or a large-bodied species, all of which tend to increase vulnerability. Human factors taken into account include the intensity of local pressure for land use, the relative costs of action in different parts of the world, and the likely extent of effective cooperation from the authorities, such as their willingness or otherwise to restrict hunting and suppress poaching. In an ideal situation, a single global strategy combining all these criteria should be established, and pursued under the auspices of an independent international body such as the IUCN.

One estimate is that to conserve 10 per cent of the present geographical range of all the living mammal species would require as much as 12 per cent of the Earth's land surface to be protected, which is perhaps easier to envisage from a conservation biologist's office in an American or European university than from that of a finance minister trying to feed a rapidly expanding, impoverished, developing nation. Thus the outlook as a whole is not good and undoubtedly a significant number, perhaps most of the 25 per cent of living mammals currently described as being at serious risk of extinction, will succumb well before the end of this century. Certainly we are not going to return to the state of 100,000 years ago in which we co-existed more or less in equilibrium with the community of mammals and all the other species in our habitat. Nor is the human population likely to fall to a level so low that a high enough proportion of the land can be given over for enough effective nature conservation areas.

Nevertheless there are a few tentative grounds for optimism. The value of nature conservation is being increasingly accepted, thanks to the ever more superb nature filming we see on our televisions, and the economic benefits arising from the demand by people who want to experience such sights and places for themselves. Precisely why conservation is so highly valued by so many people is not actually clear, since we humans could live perfectly adequate lives in the absence of all but domesticated mammals, which, to all intents and purposes, most of us actually do. Perhaps our depth of concern arises from a deep-seated, psychological affinity with the natural world within which we evolved and with which we have co-existed for most of our history—an affinity expressed nowadays mostly in our desire to keep pets, our children's cuddly toys, and our enjoyment of country walks. Perhaps it is simply an aesthetic appreciation of the patterns, colours, and scents of animals and plants. Perhaps it is the expression of a parental-like sense of responsibility for the well-being of other creatures. Perhaps the natural world offers a spiritual experience as we reflect upon our own significance in it or, for that matter, in the universe. At any event and from whatever motive, some modest successes in conservation have been achieved, and at least in theory it is clear how greater efforts might have correspondingly greater results. If we will dedicate the necessary resources, then I like to believe that at least a majority of the mammal species alive today will still be around for the next few centuries.

Further reading

Attenborough, D. (2002), *The Life of Mammals* (London: BBC Books).

Estes, R.D. (2012), *The Behaviour Guide to African Mammals* (California: California University Press).

Feldhamer, G.A., Drickamer, L.C., Vessey, S.H., Merritt, J.F., and Krajewski, C. (2015), *Mammalogy: Adaptation, Diversity, Ecology*, 4th edn (London: Johns Hopkins University Press).

Hambler, C. and Canney, S.M. (2013), *Conservation*, 2nd edn (Cambridge: Cambridge University Press).

Kemp, T.S. (2004), *The Origin and Evolution of Mammals* (Oxford: Oxford University Press).

Macdonald, D. (ed.) (2001), *The New Encyclopedia of Mammals* (Oxford: Oxford University Press).

Perrin, W., Wursig, B., and Thewissen, J.G.M. (eds) (2008), *Encyclopedia of Marine Mammals*, 2nd edn (London: Academic Press).

Petter, J.-J. (2013), *Primates of the World: An Illustrated Guide* (Princeton: Princeton University Press).

Pough, F.H., Janis, C.M., and Heiser, J.B. (2013), *Vertebrate Life*, 9th edn (Boston: Pearson).

Prothero, D.R. (2016), *The Princeton Guide to Prehistoric Mammals* (Princeton: Princeton University Press).

Savage, R.J.G. and Long, M.R. (1986), *Mammal Evolution: An Illustrated Guide* (London: British Museum (Natural History)).

Index

Mammals

Mammals

SOCIAL MEDIA
Very Short Introduction

Join our community
www.oup.com/vsi

- Join us online at the official Very Short Introductions **Facebook** page.
- Access the thoughts and musings of our authors with our online **blog**.
- Sign up for our monthly **e-newsletter** to receive information on all new titles publishing that month.
- Browse the full range of Very Short Introductions online.
- Read **extracts** from the Introductions for free.
- If you are a teacher or lecturer you can order inspection copies quickly and simply via our website.

ONLINE CATALOGUE
A Very Short Introduction

Our online catalogue is designed to make it easy to find your ideal Very Short Introduction. View the entire collection by subject area, watch author videos, read sample chapters, and download reading guides.

http://global.oup.com/uk/academic/general/vsi_list/